# Plant Hormones - Recent Advances, New Perspectives and Applications

*Edited by Christophe Hano*

Published in London, United Kingdom

IntechOpen

*Supporting open minds since 2005*

Plant Hormones - Recent Advances, New Perspectives and Applications
http://dx.doi.org/10.5772/intechopen.95608
Edited by Christophe Hano

Contributors
Musa Saheed Ibrahim, Nathan Moses, Beckley Ikhajiagbe, Nazima Rasool, Aqsa Tariq, Ambreen Ahmed, Majid Abdulhameed Ibrahim, Ehis-Eriakha Chioma Bertha, Ajuzieogu Chinyere Augusta, Akemu Stephen Eromosele, Juan F. Aguirre-Medina, Jorge Cadena-Iñiguez, Juan F. Aguirre-Cadena, Rizwan Asif, Riffat Yasmin, Madiha Mustafa, Ana Ambreen, Modasrah Mazhar, Abdul Rehman, Mukhtiar Ahmad, Shehla Umbreen, Refik Bozbuga, Bekir Bulent Arpaci, Selman Uluisik, Pakize Gok Guler, H. Nilufer Yildiz, Songul Yalcin Ates, Muhammad Amjad Bashir, Qurat-Ul-Ain Raza, Hafiz Muhammad Ali Raza, Muhammad Umair Sial, Abdur Rehim, Kashif Ali Khan, Muhammad Ijaz, Muhammad Wasif, Gunjegaonkar Shivshankar M., Wankhede Sagar B, Siraskar Balasaheb D., Merekar Abhijit N, Sachin Shinde D.



First published in London, United Kingdom, 2022 by IntechOpen
IntechOpen is the global imprint of INTECHOPEN LIMITED, registered in England and Wales, registration number: 11086078, 5 Princes Gate Court, London, SW7 2QJ, United Kingdom
Printed in Croatia

British Library Cataloguing-in-Publication Data
A catalogue record for this book is available from the British Library

Additional hard and PDF copies can be obtained from orders@intechopen.com

Plant Hormones - Recent Advances, New Perspectives and Applications
Edited by Christophe Hano
p. cm.
Print ISBN 978-1-80355-027-5
Online ISBN 978-1-80355-028-2
eBook (PDF) ISBN 978-1-80355-029-9

For EU product safety concerns:
IN TECH d.o.o., Prolaz Marije Krucifikse Kozulić 3, 51000 Rijeka, Croatia, info@intechopen.com or visit our website at intechopen.com.

# Meet the editor

Dr. Christophe Hano is a phytochemist and an assistant professor at the University of Orléans, France. His research interests include plant specialized metabolism and plant biotechnology for nutraceutical, medicinal, and cosmeceutical applications. He has published more than 200 scientific papers, reviews, and book chapters in internationally renowned journals, as well as edited one book and many journal issues.

# Contents

# Preface

Many plant hormones have been discovered since Charles and Francis Darwin initially proposed auxin activity. Their functions in the modulation of many aspects of plant growth and development (e.g., seed dormancy, seed germination, plant growth, flowering, fruit ripening, etc.) have been extensively studied since then. Plant hormones nowadays include auxins, cytokinins, gibberellins, ethylene, and abscisic acid, as well as jasmonates, salicylic acid, brassinosteroids, and strigolactones, which are essential compounds with hormone-like actions in plants. These small signaling molecules not only regulate developmental and growth activities throughout the plant life cycle, but they also initiate adaptive responses in response to external stimuli like biotic and/or abiotic stresses. Each of these plant hormones has a distinct function, yet they interact with one another in complex crosstalk, either antagonistically or synergistically. While auxins, cytokinins, gibberellins, brassinosteroids, and strigolactones have been identified as the key developmental growth-controlling plant hormones, stress regulation is commonly linked to abscisic acid, ethylene, jasmonates, and salicylic acid. In the recent decade, significant new insights into their actions during plant growth and development have been revealed. Plant hormones are also attractive candidates for modulating abiotic and biotic stresses. They are also useful for a variety of biotechnological applications. This book discusses recent advances, new perspectives, and applications of plant hormones. It is a useful resource for academics, scientists, students, and industry professionals.

**Christophe Hano**
Department of Biochemistry,
University of Orléans,
Orléans, France

## Section 1

# Phytohormones Functions

Chapter 1

# Auxins-Interkingdom Signaling Molecules

*Aqsa Tariq and Ambreen Ahmed*


## Abstract

Phytohormones play a fundamental role in the development of plants. Among various phytohormones produced by the plants, Auxins act as a master hormone that plays a major role during plant development and differentiation through cell division. Besides plants, many rhizospheric microorganisms are also capable of producing auxins specifically indole-3-acetic acid (IAA), that act as signaling molecules for the regulation of gene expressions in plants. However, bacterial IAA is majorly linked with the modulation of plant roots architecture and developing positive plant-microbe interactions. Bacterial auxin modifies root morphology by enhancing root length, forming adventitious root and root hair, thereby, increasing surface area for water and nutrient absorption affecting various aspects of plant biology in a number of ways. Bacteria mostly utilize tryptophan, present in plant root exudates, to synthesize IAA that eventually helps bacteria to colonize roots by establishing beneficial associations with plant roots. Auxins also stimulate the formation of exopolysaccharides and biofilms that help bacterial root colonization. Auxins have given the survival benefit to rhizobacteria that make them more competent to establish symbiotic interaction with plants. Synergistic and antagonistic interactions of auxins (both interkingdom and Intrakingdom) with other phytohormones play a key role in plant development and growth improvement.


**Keywords:** Auxins, bacterial IAA, phytostimulation, Indole-3-acetamide, Tryptophan

## 1. Introduction

Auxins are mainly synthesized in meristematic tissues and transported to other plant parts. Auxins play a critical role in controlling various processes during growth and development across variable environmental conditions, even at lower concentrations, these can modulate gene expression by interacting with specific proteins involved in the transcription process [1]. The plant rhizosphere is enriched with a diversity of microflora that directly contributes to their growth. The rhizosphere microbiota has the ability to produce phytohormones as a signaling molecule for inter and intraspecies communications. The synthesis and release of auxins establish a mutualistic or morbific link between organisms. Indole-3-acetic acid (IAA) is a widely produced rhizobacterial signaling phytohormone. Primarily, auxin controls various physiological processes, such as cell division, elongation, phototactic, and geotactic responses, in plants [2]. Thus, in nature, plants are receiving endogenous and exogenous signals simultaneously influencing their developmental patterns. Endogenous auxin can either be free (active auxin) or act as storage intermediates as conjugated compounds with amino acids and sugars [3]. Since higher auxin levels cause inhibitory effects, therefore, homeostasis and coordination of auxin signaling within plants and their surroundings are necessary for their regular growth and development. Endogenous auxin levels suggest the type of interactions between plant and rhizobacteria. Generally, three possible plant-bacterial IAA associations have been stated so far, first, due to direct transfer of bacterial IAA genes into host cell; second, due to bacteria living and releasing IAA within plant tissues and lastly, due to bacteria colonizing plant surfaces and producing IAA [4]. The first two associations usually result from pathogenic interactions. The knowledge of deciphering these signals and their outcomes is critical for the development of strategies for sustainable agricultural practices. Thus, the present chapter highlights the significant role of bacterial IAA as a potent microbial signaling molecule regarding beneficial plant-rhizobacterial interactions which are important for ecological resilience and sustainability.

## 2. Biosynthetic pathways of auxins

Conferring to key intermediate compounds, five different pathways for IAA synthesis have been reported in bacteria using tryptophan precursors [5]. Rhizobacteria use tryptophane either from plant root exudates or synthesize through chorismate precursor using trp gene by shikimate pathway [6, 7]. Rhizobia are an example of rhizobacteria that utilize host tryptophan for IAA synthesis [8]. Zhang *et al.* [9] analyzed 7282 prokaryotic genomes and revealed that 82.2% were efficient IAA producers from tryptophan precursors. However, Brown and Burlingham [10] observed a low amount of auxin in bacterial cultures without tryptophan indicating the fact that bacteria might have the ability to synthesize auxin without using tryptophan [11, 12]. Later, this was confirmed by the studies of Prinsen *et al.* [13] who reported the ability of IAA production by *Azospirillum brasilense* following tryptophan-independent pathway. However, there is a lack of information regarding genes, enzymes, or proteins involved. Recently, Li *et al.* [14] and Ahmad *et al.* [15] have also reported IAA biosynthesis in the absence of an exogenous tryptophan supply in Arthrobacter pascens ZZ21 and *Micrococcus aloeverae* DCB-20, respectively, however, no genetic evidence has been provided so far. Moreover, more than one auxin biosynthetic pathway functions within plants and bacteria together [9, 14].

*In vitro* production of IAA was observed to be highly influenced by bacterial growth conditions and the presence of tryptophan [16–18]. Higher auxin production by bacterial strains has been observed under increasing tryptophan concentrations [19]. Moreover, the genetic elements involved in the regulation of bacterial IAA have been demonstrated in *A. brasilense*. The key gene involved in this process is *ipdC* gene. Moreover, increased expression of *ipdC* gene was observed under increasing IAA levels indicating the involvement of auxin signaling in regulating its biosynthesis, a positive-feedback regulation. *In silico* analysis revealed that *RpoN* binding site is responsible for regulating the expression of *ipdC* gene [20]. Various transcriptional factors influencing *ipdC* gene expression have been identified in different bacterial species. Patten and Glick [21] described RpoS to regulate ipdC transcription in *Pseudomonas putida* and *P. agglomerans*, respectively. Similarly, *GacS/GacA* system has been identified in *Pseudomonas chlororaphis* as a negative regulator of tryptophan-dependent routes of IAA production [22]. Ryu and Patten [23] identified *TyrR* protein to regulate the induction of *ipdC* gene expression in Enterobacter cloacae in response to tryptophan. A high similarity of various auxin synthetic pathways has been observed between plants and bacteria with slightly different intermediate products. An overview of various auxin biosynthetic pathways has been given below:

**Indole-3-acetamide (IAM) pathway**: It involves two steps, conversion of tryptophan to Indole-3-acetamide by tryptophan-2-monooxygenase followed by conversion to IAA by IAM hydrolase [4]. The phytopathogens, such as *Agrobacterium tumefaciens*, *Pantoea agglomerans,* and *Pseudomonas syringae,* and some plant growth-promoting rhizobacterial (PGPR) genera, such as Rhizobium and Bradyrhizobium, have exhibited this pathway [7, 24].

**Indole-3-pyruvic acid (IPyA) pathway**: It involves three steps, first formation of Indole-3-pyruvic acid by aminotransferase occurs followed by decarboxylation into indole-3-acetaldehyde which is finally oxidized into IAA (**Figure 1**). The key enzyme in this pathway is identified as indole-3-pyruvate decarboxylase (encoded by ipdC gene) [4]. This pathway is present in a broad range of bacterial species from phytopathogenic bacteria (*P. agglomerans*) to PGPR (*Pseudomonas*, *Azospirillum*, *Enterobacter*, *Bacillus*, *Paenibacillus*, *Bradyrhizobium*, and *Rhizobium*) and even in cyanobacteria [7, 25].

**Tryptamine (TAM) pathway**: It involves decarboxylation of tryptophan to tryptamine which is then converted into indole-3-acetaldehyde by amine oxidase

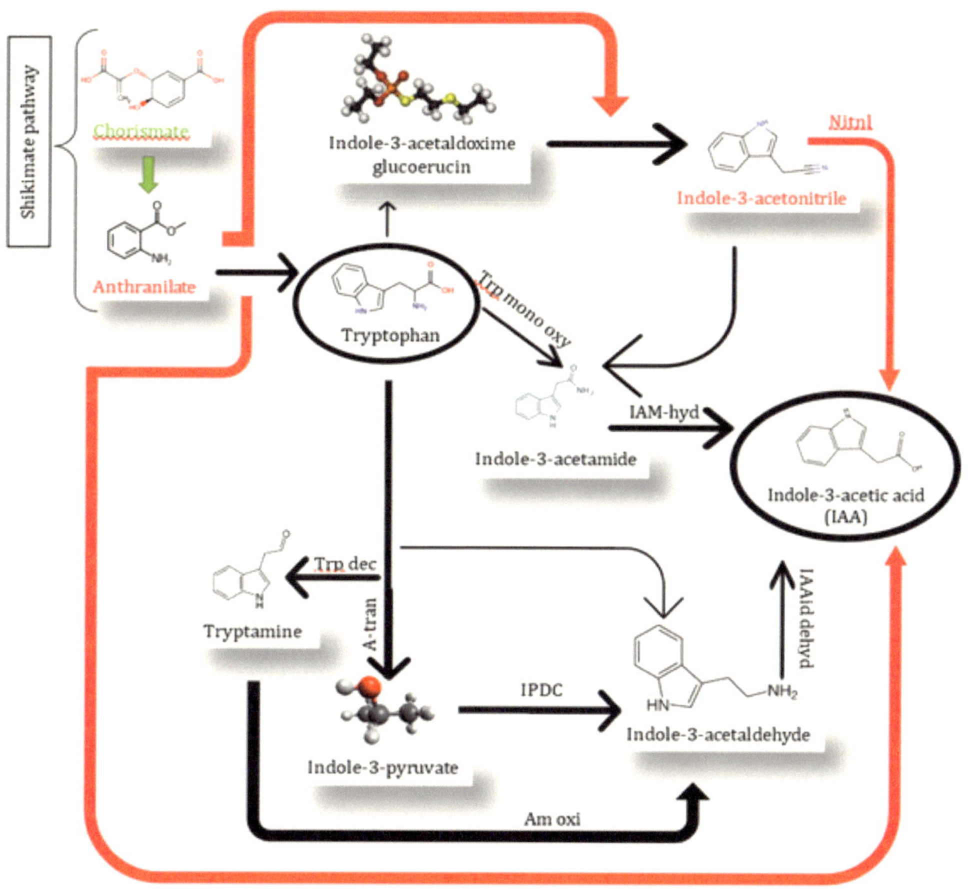


**Figure 1.**
*Various tryptophan-dependent and -independent pathways for auxin (IAA) synthesis. Red lines indicate the tryptophan-independent pathway of IAA synthesis. Black lines show tryptophan-dependent pathways. Chorismate is the precursor of both mechanisms. [A- trans - aminotransferase; Trp dec - tryptophan decarboxylase; Am oxi - amine-oxidase; IPDC - Indole-3-pyruvate decarboxylase; IAM-hyd - Indole-3-acetamide hydrolase; Nitril - Nitrilase; IAAid dehyd – Indole-3-acetaldehyde dehydrogenase].*

followed by its oxidation to IAA [4]. This has been reported in Bacillus cereus and *Azospirillum* [7, 24].

**Indole-3-acetonitrile (IAN) pathway**: In this pathway, tryptophan is converted into Indole-3-acetonitrile either by indolic glucosinolates or indole-3-acetaldoxime which is then further converted into IAA by nitrilase. This pathway has also been reported in *Alcaligenes faecalis*, *A. tumefaciens*, and *Rhizobium* spp. [7, 24].

**Tryptophan side-chain oxidase (TSO) pathway**: This is found in Pseudomonas fluorescens CHA0 and involves direct conversion of tryptophan to indole-3-acetaldehyde which then oxidizes to IAA. This mechanism is only found in bacteria and has not been studied in plants (**Figure 1**) [7, 24].

## 3. IAA – Signaling molecule

a. Intrakingdom Signaling

Auxins modulate the gene expression making it inter and intrakingdom communicating chemical messenger and quorum-sensing molecule. Scott *et al.* [26] observed bacterial chemotaxis toward IAA in *P. putida*. This movement is mediated by methyl-accepting proteins that receive and transmit IAA signals to flagellar machinery [26]. Hence, the movement of PGPR toward plant roots might be due to IAA present in root exudates. This IAA also acts as a nutrient pool, thereby, chemotaxis toward IAA ensures bacterial survival within the plant environment. Moreover, the fact that most of the plant-associated rhizobacteria produce IAA indicates that IAA might have some crucial role in bacterial cells other than interacting with plants [27]. From an evolutionary perspective, bacteria gain this ability for their survival and persistence within the plant environment [28]. IAA producers are more environmentally adaptive and competitive as compared to non-producers. Studies by Bianco *et al.* [29] showed that IAA confers protection to bacteria under various abiotic conditions, such as acidity, UV, salt, and heat stress. The author observed higher production of extracellular polysaccharides (EPS), lipopolysaccharides, and biofilms in IAA overproducers that improved bacterial adherence to plant surfaces which ultimately protect bacterial cells from various environmental stresses. Moreover, overproduction of trehalose in IAA producers has been observed indicating the accumulation of osmolytes within the bacterial cell to confer osmotic protection [29]. This was further confirmed by the studies of Donati *et al.* [30]. They reported a higher survival rate of IAA-treated bacteria under oxidative, desiccation, and osmotic stress and observed increased production of EPS and biofilm. Under various stress conditions, increased IAA levels were observed within bacterial cells indicating the fact that IAA plays important role in modulating gene expression of bacterial cells and making them more competitive [31]. However, the exact mechanism is still unknown and needed to be explored. IAA also acts as a signaling molecule for various metabolic processes within bacterial cells. Van Puyvelde *et al.* [32] observed the overall changes in gene expression of a mutant strain of *A. brasilense* and noted the decreased expression of 39 genes, including the genes involved in bacterial cell respiration by affecting the expression of NADH dehydrogenase. However, on the other hand, increased expression of the nitrate-reducing system involved in aerobic denitrification and ATP-binding cassette transporters and tripartite ATP-independent periplasmic (TRAP) transporters was also observed. Van Puyvelde *et al.* [32] also noted increased expression of T6SS (Type VI Secretion System) by exogenous IAA induction which is involved in the transport of various components via injection tube from a bacterial cell to plants cell (the mechanism by which bacteria interact directly to plant signaling pathways). Moreover, IAA also enhances the expression of genes involved in the formation of effector proteins of T3SS (Type III Secretion System) required for injection of pathogenicity within plant cells [33].

b. Interkingdom Signaling

Signal exchange between plant and rhizospheric bacteria occurs through the release of root exudates [34]. This signaling is key for developing and determining the nature of plant-bacterial interactions (symbiotic or pathogenic). PGPR colonization is the result of these signaling activities. Besides IAA synthesis, many rhizobacterial species have the ability to degrade IAA. This IAA degrading ability has given the advantage to bacteria for rhizospheric colonization and manipulating plant physiology for their survival. However, the mechanism of how IAA degradation is beneficial for plants and bacteria is not well studied and needed to be explored. Zuniga *et al*. [35] observed that IAA degradation by *Burkholderia phytofirmans* is key for efficient rhizosphere colonization and subsequent plant growth promotion. Any mutation in IAA degrading gene (iacC) also affects the growth promotional activity of the bacteria. In addition, auxin also interferes with the developmental pathways of the host. So, it is hypothesized that rhizobacteria synthesize and secrete auxin that is taken up by plants in such quantities that alter normal plant developmental pathways [36]. The principal feature of bacterial IAA reported by researchers is to manipulate the plant root growth (**Figure 2**). It induces the formation of root hair and enhances the growth of primary and lateral roots within their optimum range. However, at higher concentrations, it causes inhibitory effects and ceases the primary root growth [37]. It is suggested that this larger root system besides helping the host plant, also benefits its associated bacterial species and a larger root system absorbs more nutrients and strengthens the bacterial survival within the plant vicinity [38]. Moreover, IAA is considered to have a parallel role in developing and maintaining plant-rhizobacterial interactions [39]. In the

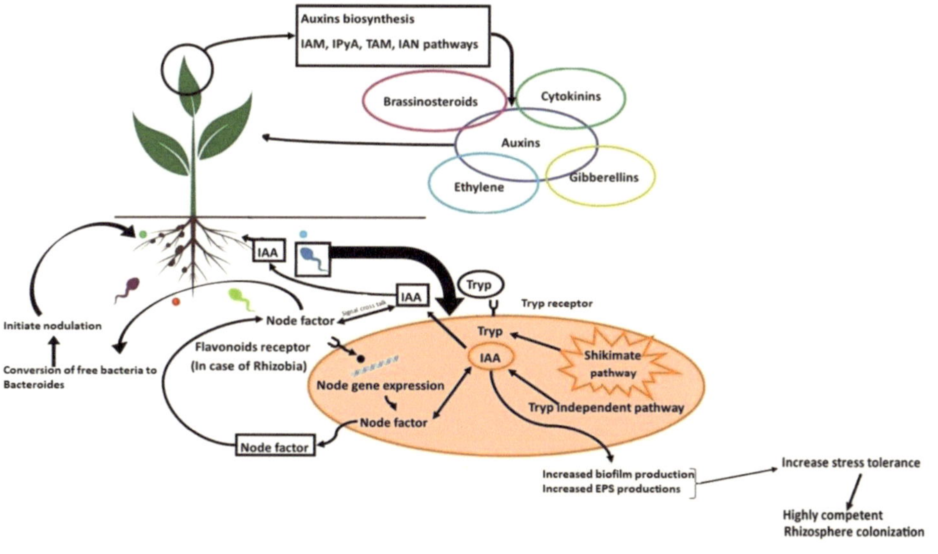


**Figure 2.**
*Interaction of root exudates to attract various auxin-producing plant beneficial bacteria leading to various metabolic activities within bacteria and making them more competent to colonize rhizosphere. Plants also uptake bacterial auxin that interacts with other phytohormones to control overall plant development.*

symbiotic association between rhizobia and legumes, the formation of macroscopic nodular structures on the roots of the host plant is considered to be formed by the action of auxin signals. Flavonoids accumulated at the sites of rhizobial entry to plant roots, inhibit auxin efflux resulting in auxin accumulation that causes excessive cell division leading to the formation of root nodules. Hence, the initiation of nodule formation is triggered by auxin signaling. Moreover, the specification of founder cells for nodule formation is also triggered by inhibition of auxin transport. Similarly, the formation of vascular bundles and the number of nodules also depend on long-distance auxin signaling. Hence, it has been hypothesized that auxin signaling triggers the formation of nodules on roots of host plants [38, 40]. Besides initiation of root nodules, IAA also modulates bacterial metabolic pathways involved in the conversion of bacteria to bacteroids for nitrogen fixation within nodules. For example, Bianco *et al.* [29] observed the activation of tricarboxylic acid and polyhydroxybutyrate cycle in *Sinorhizobium meliloti* by exogenous IAA application and in IAA-overproducer mutants (RD64). Theunis [41] observed high auxin levels in nodulated roots than in non-nodulated roots. High IAA levels also interact with nitrogen-fixing bacterial ability and enhance the nitrogen levels in nodules. In addition, the studies of Huo *et al.* [42] experimentally proven that reduction of IAA transporter genes (PIN) results in reduced nodulation. Moreover, rhizobacterial IAA also interacts with the hormonal metabolism of its associated plants. It is reported to promote the transcription of 1-aminocyclopropane-l-carboxylic acid (ACC) synthase enzyme in plants to catalyze the production of ACC deaminase enzyme which converts ACC to ammonia and α-ketobutyrate resulting in lower ethylene levels of plants. Consequently, by lowering plant ethylene levels, rhizobacteria can reduce the effect of ethylene on root growth causing plants to get nutrients and water under a wide range of stress conditions [43]. In addition, auxin signals also influence other phytohormones to regulate various plant processes. Auxin and brassinosteroids coordinate and interact to regulate the development of plant roots. Similarly, it also regulates gibberellin responses by interfering with the stability of DELLA proteins. Lower auxin levels caused reduced synthesis of gibberellins due to stabilization of DELLA proteins. Cytokinins, contrarily, have been known to suppress root formation. Therefore, overall plant growth and development depends on signaling crosstalk between auxin and other phytohormones to determine the final physiology of plants [37].

The role of bacterially produced IAA has been very significant in plant growth promotion and has been investigated by various researchers. Ahmed and Hasnain [19] studied auxin production ability and potential plant growth promotional activity of two gram-positive *Bacillus* strains and noted enhanced growth parameters, including root system and auxin content of treated plants. In another study, Fatima and Ahmed [44] investigated the role of IAA producing chromium resistant Sporosarcina saromensis and two species of Bacillus cereus on the growth of *Helianthus annuus* L. and observed an increase in plant growth parameters (shoot length, root length, fresh weight, and a number of leaves) and auxin content in treated plants. Auxin-producing bacteria stimulate seed germination and root proliferation leading to the enhanced and well-developed root system of the host plant to have greater access to water and nutrients [45]. IAA facilitates cell elongation by losing plant cell walls, thereby, increasing root length, nutrient uptake, and the release of root exudates. Enhancement in the root system of plants by exogenous application of IAA was elaborated by Vacheron *et al.* [46]. The author observed that exogenous IAA application significantly alters the root architecture of plants in a dose-dependent manner. Root growth is enhanced under optimum auxin conditions; however, higher IAA levels cease primary root growth and stimulate lateral

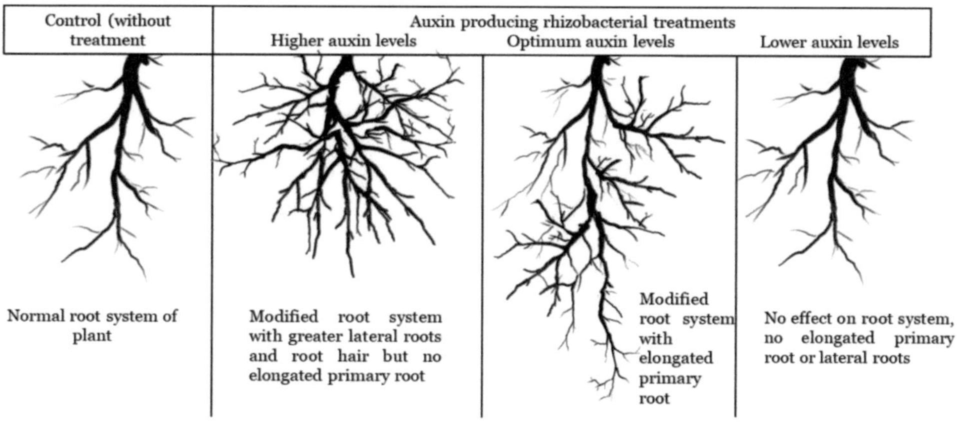


**Figure 3.**
*Root growth responses to various auxin levels.*

| Source | Plant | Effects | Cross-signaling mechanism | Reference |
|---|---|---|---|---|
| *Pseudomonas* sp. | *Zea mays* | Root architecture modified by inhibition of primary root elongation and promotion of lateral root and root hair | Auxins interacting with volatile organic compounds | [48] |
| *Streptomyces* spp. | Soybean | Enhanced shoot growth and improved dry mass | Auxins interacting with ACC deaminase production | [49] |
| *Bacillus* sp., *Agrobacterium* sp., *Rhizobium* sp., *Phyllobacterium* sp. | *Acacia cyanophylla* | Increased shoot and root dry weight, and shoot length | Auxin + solubilized phosphate interaction | [50] |
| *Bacillus* spp. | Bamboo seedlings | Enhance chlorophyll content | Auxin trigger the chlorophyll related enzymes | [51] |
| *Azospirillum brasilense* | *Arabidopsis* sp. | Increased the expression of TARGET OF RAPAMYCIN (TOR) in shoot and root apexes and induce phosphorylation | Auxin + TOR signaling | [52] |
| *Pseudomonas putida* *Pseudomonas fluorescens* | *Arabidopsis* sp. | Increased shoot and root biomass, lateral roots number per plant, and root hair formation but no effect on primary root length | Induction of auxin signaling | [47] |
| *Bacillus toyonensis* strain Bt04 | Maize roots | Enhanced root development and reduced lipid peroxidation | Induction of auxin signaling | [53] |

**Table 1.**
*Various signaling interactions of IAA.*

root growth and root hair formation. In Arabidopsis sp. greater number of lateral roots have been found in the presence of high auxin-producing *Phyllobacterium brassicacearum*, however, no effect on primary root was present. Higher levels of auxins trigger lateral root formation and initiate root hair formations. However, if the auxin concentrations in plant root do not reach optimum levels even after uptake of bacterial IAA, root growth remains unaffected. Low auxin-producing *A. brasilense* has not shown any improvement in the root growth of its associated plants [47]. Recent studies have also proven the hypothesis that bacterially produced IAA contributes toward phenotypic changes in the root architecture of treated plants (**Figure 3**) (**Table 1**) [54–56].

## 4. Conclusions

Auxin is a key phytohormone controlling the whole physiology of plants by interacting and regulating other phytohormones as well. Besides plants, various rhizobacteria have the ability to produce auxins. Various auxin biosynthetic pathways act simultaneously to regulate auxin formation. These pathways in plants and bacteria are highly similar, however, the tryptophan side chain oxidase pathway is the mechanism found only in bacteria. The main precursor for auxin synthesis is tryptophan, however, tryptophan-independent routes are also present but these routes are not well described and need to be studied. Auxin besides controlling plant growth and development, also affects various regulatory processes in bacteria as well, making inter and intrakingdom cross-signaling interactions. In bacteria, auxin primarily supports bacterial survival by strengthening their stress tolerance mechanism and also enhancing colonizing ability. This also helps in bacterial rhizospheric competence making them more adaptive to the environment. As an interkingdom signaling molecule, auxins interact with various plant signaling mechanisms and coordinate various plant growth processes. Auxins directly affect plant root architecture helping plants for enhanced nutrient and water uptake even under various stress conditions. Plants under optimum auxin levels showed enhanced and prolonged root systems but higher levels of auxin do not increase root length instead initiate the formation of lateral roots and root hair. However, very low auxin levels do not show any effect on root growth. Thus, auxins exert a significant impact (either directly or indirectly) on the healthy development and growth of the plants in a coordinated manner.

## Conflict of interest

The authors declare no conflict of interest.

## Author details

Aqsa Tariq and Ambreen Ahmed*
Institute of Botany, University of the Punjab, Quaid-e-Azam Campus, Lahore, Pakistan

*Address all correspondence to: ambreenahmed1@hotmail.com

## References


[1] Kan J, Fang R, Jia Y. Interkingdom signaling in plant-microbe interactions. Science China Life Sciences. 2017;**60**(8): 785-796. DOI: 10.1007/s11427-017-9092-3

[2] Zhao Y. Auxin biosynthesis and its role in plant development. Annu. Rev. Plant Biology. 2010;**61**:49-64. DOI: 10.1146/annurev-arplant-042809-112308

[3] Ludwig-Müller J. Auxin conjugates: Their role for plant development and in the evolution of land plants. Journal of Experimental Botany. 2011;**62**(6):1757-1773. DOI: 10.1093/jxb/erq412

[4] Patten CL, Glick BR. Bacterial biosynthesis of indole-3-acetic acid. Canadian Journal of Microbiology. 1996;**42**(3):207-220. DOI: 10.1139/M96-032

[5] Spaepen S, Vanderleyden J, Remans R. Indole-3-acetic acid in microbial and microorganism-plant signaling. FEMS Microbiology Reviews. 2007;**31**(4):425-448. DOI: 10.1111/J.1574-6976.2007.00072.x

[6] Kamilova F, Validov S, Azarova T, Mulders I, Lugtenberg B. Enrichment for enhanced competitive plant root tip colonizers selects for a new class of biocontrol bacteria. Environmental Microbiology. 2005;**7**(11):1809-1817. DOI: 10.1111/j.1462-2920.2005.00889.x

[7] Spaepen S, Vanderleyden J. Auxin and plant-microbe interactions. Cold Spring Harbor Perspectives in Biology. 2011;**3**(4):1-15. DOI: 10.1101/cshperspect. a001438

[8] Kefford NP, Brockwell J, Zwar JA. The symbiotic synthesis of auxin by legumes and nodule bacteria and its role in nodule development. Australian Journal of Biological Sciences. 1960;**13**(4):456-467. DOI: 10.1071/BI9600456

[9] Zhang P, Jin T, Sahu SK, Xu J, Shi Q, Liu H, et al. The distribution of tryptophan-dependent indole-3-acetic acid synthesis pathways in bacteria unraveled by large-scale genomic analysis. Molecules. 2019;**24**(7). DOI: 10.3390/molecules24071411

[10] Brown ME, Burlingham LSK. Production of plant growth substances by Azotobacter chroococcum. J Gen Microbial. 1968;**53**(1):135-144. DOI: 10.1099/00221287-53-1-135

[11] Merzaeva DV, Shirokikh IG. The production of auxins by the endophytic bacteria of winter rye. Applied Biochemistry and Microbiology. 2010;**46**(1):44-50. DOI: 10.1134/S0003683810010072

[12] Kiyohara S, Honda H, Shimizu N, Ejima C, Hamasaki R, Sawa S. Tryptophan auxotroph mutants suppress the superroot 2 phenotypes, modulating IAA biosynthesis in Arabidopsis. Plant Signaling & Behavior. 2011;**6**(9):1351-1355. DOI: 10.4161/psb.6.9.16321

[13] Prinsen E, Costacura A, Michiels K, Vanderleyden J, Van Onckelen H. *Azospirillum brasilense* indole-3-acetic acid biosynthesis: Evidence for a non-tryptophan dependent pathway. Molecular Plant-Microbe Interactions. 1993;**6**(5):609-615. DOI:

[14] Li M, Guo R, Yu F, Chen X, Zhao H, Li H, et al. Indole-3-acetic acid biosynthesis pathways in the plant-beneficial bacterium *Arthrobacter* pascens ZZ21. International Journal of Molecular Sciences. 2018;**19**(2):1-15. DOI: 10.3390/ijms19020443

[15] Ahmad E, Sharma SK, Sharma PK. Deciphering operation of tryptophan-independent pathway in high indole-3-acetic acid (IAA) producing Micrococcus aloeverae DCB-20. FEMS

Microbiology Letters. 2020;**24**:fnaa190. DOI: 10.1093/femsle/fnaa190

[16] Wagi S, Ahmed A. Bacillus spp.: Potent microfactories of bacterial IAA. Peer J. 2019;**7**:e7258. DOI: 10.7717/peerj.7258

[17] Ahmed A, Hasnain S. Extraction and evaluation of indole acetic acid from indigenous auxin-producing rhizosphere bacteria. J. Anim. Plant Sci. 2020;**4**:1024-1036. DOI: 10.36899/JAPS.2020.4.0117

[18] Suliasih WS. Isolation of Indole Acetic Acid (IAA) producing Bacillus siamensis from peat and optimization of the culture conditions for maximum IAA production. In IOP Conference Series: Earth and Environmental Science. 2020;**572**(1):1-12. DOI: 10.1088/1755-1315/572/1/012025

[19] Ahmed A, Hasnain S. Auxin-producing Bacillus sp.: Auxin quantification and effect on the growth of Solanum tuberosum. Pure and Applied Chemistry. 2010;**1**:313-319. DOI: 10.1351/PAC-CON-09-02-06

[20] Vande Broek A, Gysegom P, Ona O, Hendrickx N, Prinsen E, Van Impe J, et al. Transcriptional analysis of the Azospirillum brasilense indole-3-pyruvate decarboxylase gene and identification of a cis-acting sequence involved in auxin responsive expression. Molecular Plant-Microbe Interactions. 2005;**18**(4):311-323. DOI: 10.1094/MPMI-18-0311

[21] Patten CL, Glick BR. Role of Pseudomonas putida indoleacetic acid in development of the host plant root system. Applied and Environmental Microbiology. 2002;**68**(8):3795-3801. DOI: 10.1128/AEM.68.8.3795-3801.2002

[22] Kang BR, Yang K, Y, Cho B. H, Han T. H, Kim I. S, Lee M. C, Anderson A. J, Kim Y. C., Production of indole-3- acetic acid in the plant-beneficial strain Pseudomonas chlororaphis O6 is negatively regulated by the global sensor kinase Gac S. Current Microbiol. 2006;**52**:473-476. DOI: 10.1007/s00284-005-0427-x

[23] Ryu RJ, Patten CL. Aromatic amino acid-dependent expression of indole-3-pyruvate decarboxylase is regulated by TyrR in Enterobacter cloacae UW5. J Bacteriology. 2008;**190**(21):7200-7208. DOI: 109.1128/JB.00804-08

[24] Keswani C, Singh SP, Cueto L, García-Estrada C, Mezaache-Aichour S, Glare TR, et al. Auxins of microbial origin and their use in agriculture. Applied Microbiology and Biotechnology. 2020;**140**:1-17. DOI: 10.1007/s00253-020-10890-8

[25] Persello-Cartieaux F, Nussaume L, Robaglia C. Tales from the underground: molecular plant–rhizobacteria interactions. Plant Cell Environment. 2003;**26**(2):189-199. DOI: 10.1046/j.1365-3040.2003. 00956.x

[26] Scott JC, Greenhut IV, Leveau JHJ. Functional characterization of the bacterial iac genes for degradation of the plant hormone indole 3-acetic acid. Journal of Chemical Ecology. 2013;**39**: 942-951. DOI: 10.1007/s10886-013-0324-x

[27] Patten CL, Blakney AJC, Coulson TJD. Activity, distribution and function of indole-3-acetic acid biosynthetic pathways in bacteria. Critical Reviews in Microbiology. 2013;**39**(4):395-415. DOI: 10.3109/1040841X.2012.716819

[28] Kim JI, Murphy AS, Baek D, Lee S, Yun D, Bressan RA, et al. L (2011a) YUCCA6 over-expression demonstrates auxin function in delaying leaf senescence in Arabidopsis thaliana. Journal of Experimental Botany. 2011;**62**(11):3981-3992. DOI: 10.1093/jxb/err094

[29] Bianco C, Imperlini E, Defez R. Legumes like more IAA. Plant Signaling & Behavior. 2009;**4**:763-765. DOI: 10.4161/psb.4.8.9166

[30] Donati AJ, Lee HI, Leveau JHJ, Chang WS. Effects of indole-3- acetic acid on the transcriptional activities and stress tolerance of Bradyrhizobium japonicum. PLoS One. 2013;**8**:1-11. DOI: 10.1371/journal.pone.0076559

[31] Molina R, Rivera D, Mora V, López G, Rosas S, Spaepen S, et al. Regulation of IAA biosynthesis in Azospirillum brasilense under environmental stress conditions. Current Microbiology. 2018;**75**(10):1408-1418. DOI: 10.1007/s00284-018-1537-6

[32] Van Puyvelde S, Cloots L, Engelen K, Das F, Marchal K, Vanderleyden J, et al. Transcriptome analysis of the rhizosphere bacterium Azospirillum brasilense reveals an extensive auxin response. Microb Ecology. 2011;**61**:723-728. DOI: 10.1007/s00248-011-9819-6

[33] Yang S, Zhang Q, Guo J, Charkowski AO, Glick B. R, Ibekwe A. M. *et al*. Global effect of indole-3-acetic acid biosynthesis on multiple virulence factors of Erwinia chrysanthemi 3937. Appl Environ Microbiology, 2007; 73(4):1079–1088. DOI:10.1128/AEM.01770-06

[34] Phour M, Sehrawat A, Sindhu SS, Glick BR. Interkingdom signaling in plant-rhizomicrobiome interactions for sustainable agriculture. Microbiological Research. 2020;**241**. DOI: 10.1016/j.micres.2020.126589

[35] Zuniga A, Poupin MJ, Donoso R, Ledger T, Guiliani N. Quorum sensing and indole-3-acetic acid degradation play a role in colonization and plant growth promotion of Arabidopsis thaliana by Burkholderia phytofirmans PsJN. Molecular Plant-Microbe Interactions. 2013;**26**(5):546-553. DOI: 10.1094/MPMI-10-12-0241-R

[36] Sukumar P, Legue V, Vayssieres A, Martin F, Tuskan GA, Kalluri UC. Involvement of auxin pathways in modulating root architecture during beneficial plant–microorganism interactions. Plant, Cell & Environment. 2013;**36**(5):909-919. DOI: 10.1111/pce.12036

[37] Ahmed A, Hasnain S. Auxins as one of the factors of plant growth improvement by plant growth promoting rhizobacteria. Polish Journal of Microbiology. 2014;**3**:261-266

[38] Ludwig-Müller J. Auxin and the interaction between plants and microorganisms. Auxin and its role in plant development. 2014:413-434. DOI: 10.1007/978-3-7091-1526-8_18

[39] Duca DR, Glick BR. Indole-3-acetic acid biosynthesis and its regulation in plant-associated bacteria. Applied Microbiology and Biotechnology. 2020: 1-13. DOI: 10.1007/s00253-020-10869-5

[40] Mathesius U. Auxin: at the root of nodule development? Functional Plant Biology. 2008;**35**:651-668. DOI: 10.1071/FP08177

[41] Theunis M. IAA biosynthesis in rhizobia its potential role in symbiosis (thesis). Universiteit Antwerpen, Antwerp, Belgium; 2005

[42] Huo X, Schnabel E, Hughes K, Frugoli J. RNAi phenotypes and the localization of a protein: GUS fusion imply a role for Medicago truncatula PIN genes in nodulation. J Plant Growth Reg. 2006;**25**:156-165. DOI: 10.1007/s00344-005-0106-y

[43] Glick BR. Bacteria with ACC deaminase can promote plant growth and help to feed the world. Microbiological Research. 2013;**169**(1):30-39. DOI: 10.1016/j.micres.2013.09.009.0

[44] Fatima H, Ahmed A. Indole-3-acetic acid synthesizing chromium-resistant bacteria can mitigate chromium toxicity in Helianthus annuus L. Plant, Soil and Environ. 2020;**66**(5):216-221. DOI: 10.17221/581/2019-PSE

[45] Ghosh A, Pramanik K, Bhattacharya S, Mondal S, Ghosh SK,

Ghosh PK, et al. Abatement of arsenic-induced phytotoxic effects in rice seedlings by an arsenic-resistant Pantoea dispersa strain. Environmental Science and Pollution Research. 2021;**28**(17):1-17. DOI: 10.1007/s11356-020-11816-7

[46] Vacheron J, Desbrosses G, Bouffaud ML, Touraine B, Moenne-Loccoz Y, Muller D, et al. Plant growth-promoting rhizobacteria and root system functioning. Frontiers in Plant Science. 2013;**4**:356. DOI: 10.3389/fpls.2013.00356

[47] Ortiz-Castro R, Campos-García J, López-Bucio J. Pseudomonas putida and Pseudomonas fluorescens influence Arabidopsis root system architecture through an auxin response mediated by bioactive cyclodipeptides. Journal of Plant Growth Reg. 2020;**39**(1):254-265. DOI: 10.1007/s00344-019-09979-w

[48] Chu TN, Bui LV, Hoang MTT. Pseudomonas PS01 isolated from maize rhizosphere alters root system architecture and promotes plant growth. Microorganisms. 2020;**8**(4):471. DOI: 10.3390/microorganisms8040471

[49] Horstmann JL, Dias MP, Ortolan F, Medina-Silva R, Astarita LV, Santarém ER, et al. CLV45 from Fabaceae rhizosphere benefits growth of soybean plants. Brazilian Journal of Microbiology. 2020;**51**(4):1861-1871. DOI: 10.1007/s42770-020-00301-5

[50] Lebrazi S, Niehaus K, Bednarz H, Fadil M, Chraibi M, Fikri-Benbrahim K. Screening and optimization of indole-3-acetic acid production and phosphate solubilization by rhizobacterial strains isolated from Acacia cyanophylla root nodules and their effects on its plant growth. Journal of Genetic Engineering and Biotechnology. 2020;**18**(1):1-12. DOI: 10.1186/s43141-020-00090-2

[51] Maya KCB, Gauchan DP, Khanal SN, Chimouriya S, Lamichhane J. Extraction of Indole-3-acetic Acid from Plant Growth Promoting Rhizobacteria of Bamboo Rhizosphere and Its Effect on Biosynthesis of Chlorophyll in Bamboo Seedlings. Indian Journal of Agricultural Research. 2020;**54**(6):1-7. DOI: 10.18805/IJARe.A-5578

[52] Méndez-Gómez M, Castro-Mercado E, Peña-Uribe CA, Reyes-de la Cruz H, López-Bucio J, García-Pineda E. TARGET OF RAPAMYCIN signaling plays a role in Arabidopsis growth promotion by Azospirillum brasilense Sp245. Plant Science. 2020;**293**(110416). DOI: 10.1016/j.plantsci.2020.110416

[53] Zerrouk IZ, Rahmoune B, Auer S, Rößler S, Lin T, Baluska F, et al. Growth and aluminum tolerance of maize roots mediated by auxin-and cytokinin-producing Bacillus toyonensis requires polar auxin transport. Environmental and Experimental Botany. 2020;**176**: 104064

[54] Jochum MD, McWilliams KL, Borrego EJ, Kolomiets MV, Niu G, Pierson EA, et al. Bioprospecting plant growth-promoting rhizobacteria that mitigate drought stress in grasses. Frontiers in Microbiology. 2019;**10**:2106. DOI: 10.3389/fmicb.2019.02106

[55] Batista BD, Dourado MN, Figueredo EF, Hortencio RO, Marques JPR, Piotto FA, et al. C., The auxin-producing Bacillus thuringiensis RZ2MS9 promotes the growth and modifies the root architecture of tomato (Solanum lycopersicum cv. Micro-Tom). Archives of Microbiology. 2021;**23**:3869-3882

[56] Oliveira DAD, Ferreira SDC, Carrera DLR, Serrao CP, Callegari DM, Barros NLF, et al. Characterization of Pseudomonas bacteria of Piper tuberculatum regarding the production of potentially bio-stimulating compounds for plant growth. Acta Amazonica. 2021;**51**:10-19

Chapter 2

# Plant Hormones: Role in Alleviating Biotic Stress

*Nazima Rasool*

## Abstract

Plant hormones play a critical role in regulating plant developmental processes. Jasmonic acid, salicylic acid and brassinosteroids have been recently added to the list of plant hormones apart from auxins, gibberellins, cytokinins, abscisic acid and volatile hormone ethylene. Besides their regulatory role in plant development, plant hormones, ethylene, Jasmonic acid and salicylic acid play key roles in the plant defense response while as auxins, gibberellins, abscisic acid, cytokinins and brassinosteroids are known to modulate their effects. For an effective response to biotic stresses, the signaling pathways of different hormones are integrated at different levels enabling crosstalk between them. In this chapter, I will analyze how plant hormones signal defense response and interact with each other through crosstalk to regulate plant defense.

**Keywords:** plant hormones, disease response, biotic stress, hormone cross-talk

## 1. Introduction

Plant productivity is threatened by biotic and abiotic stress. In order to feed the world population of over 7 billion at the moment, productivity needs to be safeguarded against biotic and abiotic stresses. Biotic stress is caused due to attacks of viruses, bacteria, fungi, nematodes and other pathogens and pests. Pathogens are usually categorized into biotrophs and necrotrophs. Although the former penetrate the epidermal cells, multiply inside the intercellular spaces and feed on the living host tissue the latter kill the host cells and then feed on the cell remains. Biotrophs are mostly host-specific, the nectrophs have a broader host range [1]. Agricultural intensification has already led to increased soil pollution and land degradation problems. Therefore, understanding the natural mechanisms of defense in plants against various kinds of stresses is important to exploit it in a sustainable and environment-friendly manner. Of the various mechanisms plants have developed to combat biotic stress, hormones are of primary importance. Plant hormones are biochemicals that are synthesized at one location in plants and bring about the desired effect at the same or different location, at unimaginably low concentrations. Plant hormones are diverse in their chemical nature and biological functions derived from amino acids (IAA, ethylene), lipids (Jasmonic acid), from the isoprenoid (cytokinins, gibberellins, abscisic acid etc) and chorismate (salicylic acid) pathways (**Figure 1**). There are many biomolecules that have been added to the list of plant or phytohormones of late, which include jasmonic acid (JA), salicylic acid (SA), strigolactones (SL), brassinosteroids (BR) and peptides, besides auxins (IAA), gibberellins (GA), abscisic acid (ABA), cytokinins (CK) and ethylene (ET) that have been there since a long time. Salicylic acid, jasmonic acid and ethylene play very important roles

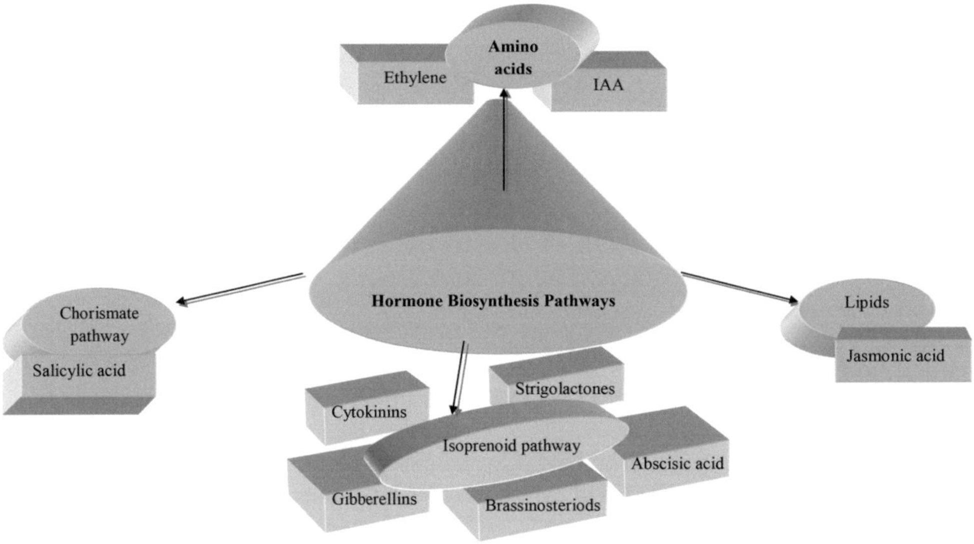


**Figure 1.**
*Pathways of hormone biosynthesis.*

| Hormone | Nature | Stress | Mechanism | References |
|---|---|---|---|---|
| Ethylene | Alkene | Biotic Stress (Necrotrophic pathogens and herbivores) | ISR; Interactions with JA; modulates JA/SA antagonism; induces defense genes such as *Plant Defensin1.2* (*PDF1.2*) or *THI2.1* (thionin), *Vegetative Storage Protein2* etc. | [3] |
| Jasmonic acid | Sesquiterpene | Biotic stress | Interacts with many other hormones to mediate stress response; ISR | [4, 5] |
| | | Abiotic stress | It activates the antioxidant system, causes accumulation of amino acids, and soluble sugars and regulates stomatal opening and closing | [6] |
| Salicylic acid | Phenolic compound | Abiotic stress | Increase antioxidant activity | [7] |
| | | Biotic stress | JA/ SA antagonism perfects pathogen specific response; mediates SAR; also involved together with MAPK signaling in resistance to aphids | [3, 8, 9] |
| Gibberellins | Diterpenoid | Abiotic stress | Degradation of DELLAs | [10] |
| | | Biotic stress Reverse the inhibitory effect of different stress conditions in seed germination and seedling establishment | Affect the relative strength of SA/JA signaling; Induction of salicylic acid (SA)-dependent defense pathway | [3, 10, 11] |
| | | Modulate SA Biosynthesis | Through degradation of DELLAs | [10] |

| Hormone | Nature | Stress | Mechanism | References |
|---|---|---|---|---|
| Auxins | Tryptophan derivative | Drought stress | By modulating root architecture, ABA (abscisic acid)-responsive genes expression, and ROS metabolism | [12] |
| | | Biotic stress | Contributes to SAR | [8] |
| | | | Interaction with other hormones | [13] |
| Abscisic acid | Isoprenoid | Biotic stress | Influences the Central backbone (SA-JA/ET) of plant defense | [3] |
| | | Abiotic stress | Stomatal closure; reduction in ROS levels | [3] |
| Cytokinins | Isopentenyladenine derivative | Biotic stress | Through direct interactions of CK-signaling components with the Central phytohormonal immunity backbone; interact with SA to induce defense responses via WRKY45 and NPR1; Phytoalexin accumulation | [3, 14] |

**Table 1.**
*Role of various plants hormones in biotic and abiotic stress response.*

in plant biotic stress response [2] while as, auxins, abscisic acid and gibberellins etc. modulate it. An overview of the important roles of major plant hormones is presented in **Table 1**. The role of hormones in plant growth and development is largely known and mechanisms of their biosynthesis have been elucidated in the majority of the cases, what remains to be fully understood is their mediation of the defense response in plants. In this chapter, I discuss how these hormones mediate the plant defense response and also assess how their effects are modulated by other hormones.

## 2. The major players in the plant defense

**Ethylene (ET)**: This is a gaseous hormone that is responsible for various functions in plants notably fruit ripening, flower senescence and abscission of leaves etc. In dark-grown seedlings ET causes inhibition of hypocotyls and root elongation, radial swelling of hypocotyls and exaggeration of the apical hook this is commonly called as the triple response [15, 16]. The role of ET in plant stress is very well known [17–19], it favors stress resistance over growth thereby increasing stress tolerance [20–23]. ET acts in cooperation with JA to present an effective defense response against necrotrophic and chewing insects [1, 24]. ET response involves a battery of ET receptors which, for example, in *Arabidopsis* (*Arabidopsis thaliana*) include ETHYLENE RESPONSE 1 (ETR1), ETR2, ETHYLENE RESPONSE SENSOR 1 (ERS1), ERS2 and ETHYLENE INSENSITIVE 4 (EIN4) [24–28]. Mutations in these lead to ET insensitivity and increased susceptibility to the necrotrophic pathogens [24, 29, 30]. ein3 and eil1 (ETHYLENE INSENSITIVE LIKE) double mutants are completely ethylene insensitive and these lacks the triple response, pathogen resistance and the ability to fully suppress ctr1 mutation [15, 31, 32]. CONSTITUTIVE TRIPLE RESPONSE (CTR) is the negative regulator of the ET pathway in absence of ET. When ET binds to ER anchored EIN2-protein receptors, it causes dephosphorylation of the latter. This leads to cleavage of the C-terminal domain (EIN2-C) of the EIN2 [1, 15, 33, 34]. EIN2-C moves to the nucleus and triggers EIN3 and EIN3-like transcription factors, eliciting ET-mediated response [15]. Transcription of *ETHYLENE INSENSITIVE3* (EIN3) activates defense response by inducing expression of *ERF1* [19, 35]. Repression of EIN2 by CONSTITUTIVE TRIPLE RESPONSE (CTR1) is released after perception of ET by ETHYLENE RESPONSE 1 (ETR1) [19, 36]. ET release stabilizes EIN3/EIL1 levels [15]. At the same time, ET also decreases levels of EBF1/2 (EIN3 BINDING F-BOX 1) protein through suppression of translation of its mRNA in the cytosol promoted by EIN2-C [15, 37, 38] and by EIN2-dependent proteasomal degradation of EBF1/2 proteins [15, 39]. EIN3 leads to EBF1/2 expression providing a negative feedback loop to the ET signaling [1].

**Jasmonic acid (JA):** Methyl jasmonate and its free acid jasmonic acid are collectively known as jasmonates. Jasmonic acid is the better known of the two while JA-Ile is the active form [1]. JA is a cyclopentane fatty acid that is synthesized from linolenic acid which is a common constituent of plant cell membranes [40]. JAs play a vital role in the various plant developmental processes including flowering, fruiting, senescence and secondary metabolism. These are known to be critically important in plant defense and abiotic stress response [41–43]. JA activates the antioxidant system, causes the accumulation of amino acids, and soluble sugars and regulates the stomatal opening and closing during abiotic stress [6, 44]. JA interacts with SA, ET and ABA during plant defense response [1, 4, 45]; its interactions with auxins, gibberellins and cytokinins during important development processes like root, stamen, hypocotyl, xylem development etc. are also well known [40]. The JAs affect both plant development and plant stress-resistance [46, 47].

Coronatine Insensitive Receptor (COI1) and JAZ (Jasmonate-ZIM domain) proteins mediate JA-signaling pathway [24, 48, 49]. The others involved in JA signaling include JASMONATE INSENSITIVE 1/MYC2 (JIN1/MYC2) and several members of the APETALA2/ETHYLENE-RESPONSIVE FACTOR (AP2/ERF) family [19, 50]. COI1 forms a part of the E3 –ubiquitin degradation complex as SCF $^{COI1}$ complex. The SCF consists of Skp-1/Cullin/F-box. JAZ is a repressor of the JA response. SCF $^{COI1}$ complex binds with JAZ repressors at higher JA concentrations and this leads to ubiquitination and degradation of JAZ mediated by 26S proteasome [1]. The JA signaling pathway may follow two paths one is the MYC pathway and another is the ERF pathway. Wounding and insect feeding induces the MYC branch which further involves MYC2, MYC3 and MYC4 - basic helix–loop–helix leucine zipper transcription factors [1]. In absence of JA-Ile JAZ proteins interact with JIN1/MYC2 and inhibit transcriptional regulation of JA-responsive genes [19]. Interaction of JAZ with MYC proteins competitively inhibits their interaction with the MED25 subunit of the Transcriptional Mediator Complex [1, 51]. This causes the expression of several JA responsive genes including VSP2 (vegetative storage protein), JA synthesis gene LOX2 and JA signaling repressor JAZ genes. ERF pathway is stimulated by necrotrophic pathogens. As the name indicates this branch is regulated by ET; AP2/ERF-domain transcription factors ORA59 and ERF (ERF1, ERF2, ERF5 and ERF6) control this branch. ORA59 and ERF1 bind to GCC-box motif through ERF domain and activate the expression of PDF1.2 which is the marker gene of this pathway [1, 19, 52–55]. The mode of interaction between JAZ and ERFs is not known. EIN3 directly interacts with JAZ which represses the expression of *ORA59* and *ERF1* [1, 56]. Given the two branches MYC and ERF are induced under different kinds of pathogen attacks these two are mutually antagonistic [1].

Coi1 mutants lacking JA response are more susceptible to necrotrophic pathogens including *Botrytis cinerea*, *Pythium irregulare*, *Alternaria brassicicola* and other pathogens [24, 57–60]. Susceptibility to herbivores is increased by mutations stabilizing JA e.g., JAZ1Δ3A mutation increases susceptibility to *Spodoptera exigua* [24, 61]. The fine-tuning of JA-mediated defense response mediated via MYC2 is achieved by post-translational phosphorylation at thr328 residue that makes it unstable leading to its degradation by plant Ubox protein (PUB10) that works as E3 ligase facilitating MYC2 turn over [19, 62].

**Salicylic acid (SA):** this is a phenolic acid hormone that plays important role in the regulation of plant growth, fruit ripening and development. It is involved in pathogenesis-related protein expression [63, 64]. It may be synthesized through the shikimic acid pathway either via the isochorismate branch or phenylalanine ammonia-lyase branch. Salicylic acid regulates the expression of genes encoding molecular chaperones, heat shock proteins, antioxidants and those involved in the biosynthesis of secondary metabolites, alcohol dehydrogenases and cytochrome P450 [64, 65]. In recent years, SA has been increasingly implicated in the plant defense response [10, 66]. Increased SA biosynthesis improves plant tolerance to salt, oxidative and heat stress [10] and it is synthesized in response to pathogen attack [19]. Meaning thereby it has a role to play in biotic as well as the abiotic stress response. SA also leads to systemic acquired resistance (SAR) - defense response to a secondary pathogen infection far and wide in the plant after it has been exposed to a pathogen previously [19]. SA is accumulated in the plant tissue before SAR is initiated [24]. During SAR there is oxidative burst which is followed by increased levels of antioxidants to neutralize the harmful effects of the reactive oxygen species [24]. Mutations in SA-related genes compromise plant immunity to pathogens and diminish the expression of anti-microbial proteins [19, 67]. sid2–1 *Arabidopsis* mutants with impaired SA biosynthesis show reduced pathogen resistance [68]. In transgenic *Arabidopsis* plants expressing bacterial SA hydroxylase gene nahG, which causes the conversion of SA to catechol, SAR is not activated instead PR gene expression is activated [24, 69, 70]. SA effects SAR by affecting the expression of various genes including PAL and priming genes, it activates phytoalexin and auxin signaling-related pathways, it also effects the deposition of callose and phenolic products [46, 71]. SA induces resistance against biotrophic and hemibiotrophic pathogens including *Hyaloperonospora arabidopsidis* and *Pseudomonas syringae*.

SA signaling involves NPR1 (non-expressor of pathogenesis-related (PR) genes, a protein with ankyrin repeat [8, 72]. NPR1 is an oligomer formed by intramolecular disulfide bridges under uninduced conditions [8, 73]. SA induces de-oligomerization of NPR1 releasing active monomers which migrate to the cell's nucleus inducing expression of PR genes [8, 74]; only the monomeric forms can interact with the TGA (TGACG binding) transcription factors which are bZIP proteins [8, 19, 75]. This facilitates the binding of TGA transcription factors with promoters of NPR1 dependent genes [8, 76]. The triple mutant tga2 tga5 tga6 does not respond to SA and does not have SAR [8, 77]. Both NPR1 and TGA undergo nitrosylation which increases the DNA binding ability of the latter. Thiol S-nitrosylation on the other hand causes oligomerization of the NPR1 leading to its inactivation [8, 74]. NPR1 undergoes phosphorylation and proteasome-degradation thereby allowing its turnover [78]. NRR3/4 also interacts with TGA and mutants nrp3/4 over accumulate NPR1 leading to faulty SAR [8]. The binding of NPR3 and NPR4 with Cullin 3 ubiquitin E3 ligase causes SA-dependent NPR1 degradation [8, 79]. The binding of NPR with SA causes a conformational change in the NPR1 required for NPR1 dependent PR gene expression. NPR is also important in epigenetic effect-dependent trans-generational immunity in plants [8, 80]. Pathogen resistance in Monocots is enhanced by over expression of NPR1 [8, 81].

## 3. The modulators of the plant defense

**Auxins (IAA):** The role auxins play in plant growth and development is very well known however, their involvement in plants' response to the biotic stress has only begun to be elucidated [64, 82]. Auxins control apical dominance, tropic responses, development of vascular cambium, organ patterning, flower and fruit development [13]. Auxin/indole acetic acid (Aux/IAA), Auxin response factors (ARF), TOPLESS (TL) proteins are the transcriptional regulators that affect cell-specific transcription of auxins and are involved in auxin signaling [83–87]. Research on ARF has led to their identification and characterization from several pants including *Arabidopsis* [88], maize [89], rice [85, 90], poplar [91], tomato [87, 92, 93], Chinese cabbage [94], sorghum [95] and banana [83, 96].

Important auxin-responsive genes include *Aux/IAA*, *GH3* and *SAUR* gene families. In a study by Ghanshyam and Jain [13], 154 auxin-induced and 161 auxin-repressed genes were reported to express differentially under the biotic stress-induced by *Magnaporthe grisea* and *Striga hermonthica*. 62 of the auxin-induced genes were common to both the pathogens while others showed a specific response, 55 to *M. grisea* and 37 to *S. hermonthica*. In the category of auxin-repressed genes, 16 genes showed response to both the pathogens while others showed a specific response, 10 to *M. grisea* and 35 to *S. hermonthica*. Altered expression in auxin genes has also been reported in cotton in response to *Fusarium oxysporum* f. sp. vasinfectum infection [13, 97]. *Botrytis cinerea* infection in arabidopsis causes down-regulation of all the auxin-responsive genes [98]. Repression of auxin-mediated signaling through micro-RNAs leads to resistance against *P. syringae* in arabidopsis. Various pathogens operate by modulating auxin levels *in planta* to enhance the host susceptibility [99, 100]. Exogenous IAA increased susceptibility to *Xanthomonas oryzae* pv. oryzae due to cell wall loosening effects of auxins [101]. The *P. syringae* type III effector AvrRpt2 causes altered auxin levels and modified auxin-related phenotypes and decreases resistance against Pst DC3000 in *Arabidopsis* plants lacking expression of *RPS2* [24, 102]. In this case, susceptibility was found to be directly related to the auxin levels. Another study conducted by Naseem et al. [103], also supports that auxins, JA and ABA increase the host's susceptibility. Naseem and Dandekar, [104] propose a working model of the interaction between auxins and CK stating that the pathogens increase auxin levels increasing disease susceptibility by decreasing SA- and CK-based disease response while as CK pretreatment influences auxin synthesis and transport thereby increasing resistance [24].

Auxins down-regulate jasmonic acid biosynthesis genes in *Arabidopsis* [13, 105]. GH3.5 acts as a bifunctional modulator in auxin and SA signaling [106]. Overexpression of *GH3.5* leads to accumulation of SA and accumulation of pathogenesis-related −1 gene (PR-1 gene) product while the capacity of systemic acquired resistance (SAR) is compromised in gh3.5 mutants [106]. Resistance of arabidopsis against *X. oryzae* is increased due to the over expression of *GH3–8* [101].

**Abscisic acid (ABA):** ABA is a sesquiterpene that is synthesized from carotenoids [107]. ABA is usually known as "stress hormone", it regulates a wide range of processes to increase a plant's stress tolerance [23, 108, 109]. It is known to influence the expression of 10% of the protein-coding genes in the events of stress [64, [110]. Important components in ABA signaling include PYR/PYL/RCAR etc. [107]. This hormone oversees important functions in plants that include, seed dormancy, accumulation of nutrient reserves in the developing seeds, desiccation tolerance and arrest of embryonic development during seed maturation [111]. It also plays important role in the protein synthesis and synthesis of some osmolytes [112]. ABA may positively or negatively modulate defense response depending on the type of the pathogen [113–116]. Impaired biosynthesis or signaling ABA mutants in *Arabidopsis* (*abi1–1*, *abi2–1*, *aba1–6*, *aba2–12*, *aao3–2*, and *pyr1pyl1pyl2pyl4*) and tomato (*sitiens*) showed increased resistance to *B. cinerea*, *P. syringae*, *Fusarium oxysporum*, *Plectosphaerella cucumerina* and *Hyaloperonospora parasitica* [117–122]. Antagonistic interactions between ABA and major plant defense hormones including ET, JA and SA have been reported and it has been found that 65% of the genes upregulated and 30% of the genes down-regulated in aba1–6 mutants are those that are affected (up- or down-regulated) by ET, JA or SA treatment [119, 122, 123]. The genes constitutively up/down-regulated in these mutants were also found to differentially express upon infection by *P. cucumerina*, indicating their role in the defense. ABA also plays role in expression of R genes [124]. ABA deficient plants are more susceptible than wild types to pathogens *Alternaria brassicicola*, *Ralstonia solanacearum* and *Pythium irregulare* [57, 125, 126]. JA biosynthesis needs ABA in *Arabidopsis* for *P. irregulare* resistance [57]. However, negative interaction between the two is known in the case of the *P. cucumerina* [122]. It has been reported that inoculation of *Arabidopsis* with avirulent strains of *R. solanacearum* makes the plant resistant to the virulent strains of the bacterium and the resistance is mediated through ABA, hence, this hormone could be used for controlling wilt induced by this pathogen [114].

**Cytokinins (CK)**: The most important aspect of cytokinin function in plants is the maintenance of the identity of stem cells thus cytokinins affect the basic aspects of the growth and development of plants [40, 127]. CK was first identified as a hormone affecting cell division in tissue culture conditions and now its role in regulating the cell cycle is well known [23, 128]. Various functions regulated by cytokinins include inhibition of lateral root initiation and leaf senescence [23, 129–132] differentiation of vascular tissue (phloem and metaxylem in roots [23, 133, 134], morphogenic differentiation in expanding leaves and regulation of their cell division [23, 135–137] Cytokinins are derivatives of isopentenyladenine; zeatin is a common CK [40, 127, 138] which exists in two forms -cis and trans-zeatin. Trans-zeatin is more active [40, 139, 140] and is produced by isopentenyl transferases and cytochrome P450CYP735A1 and CYP735A2 [40, 127]. Activity and homeostasis of cytokinins are regulated by their degradation or conjugation with glucose and amino acids, and CK oxidase which cleaves cytokinins [40, 139, 140].

The role of cytokinins in plant defense was first recorded from tobacco plants with down-regulated S-adenosyl homocysteine hydrolases; the plants had higher resistance to tobacco mosaic virus, cucumber mosaic virus, potato virus X, and potato virus Y and also showed increased levels of CK and higher levels CK-related developmental defects [24, 141]. Cytokinin deficient plants have higher stress tolerance [40, 43, 142–144]. Several cytokinin receptors, histidine phosphotransfer proteins and transcription factors mediate CK signaling. Three histidine kinases (AHK2, AHK3, and AHK4/WOODEN LEG) working as cytokinin receptors have been identified in *Arabidopsis* [127, 145]. Cytokinins cause autophosphorylation of the conserved histidine residues in these kinases [40]. The phosphate is transferred to the histidine phosphotransfer proteins (AHPs) through aspartate residue. Phosphorylated AHPs move to the cellular nucleus activating B-TYPE ARABIDOPSIS RESPONSE REGULATOR (ARR) transcription factors causing transcription of cytokinin response genes [146, 147]. Both the environmental factors as well as the JA levels in plants affect the components of the cytokinin response system [40, 148–151]. It is believed that JA may be controlling CK response through MYC2 by promoting AHP expression [40]. Thus interactions between JA and CK might occur at the levels of signaling response elements [40]. Many studies indicate that CKs affect the plant defense response mediated by SA and JA. CK is believed to affect priming in SAR and affects the synthesis of SA and PR proteins [24, 152–154]. Exogenous supply and internal increased levels of CK increase the JA levels to hasten the defense reaction in wounded plants [24, 155, 156]. The mechanism employed by CK for disease protection is different in different plants e.g., in solanaceous plants, it increases the ratio of phytoalexin to pathogen restricting pathogen development [24]. JA accumulation increases CK ribosides in potato [24, 157]. Several genes involved in regulating CK levels in plants including IPT and CKX are seen as purported targets in enhancing plant disease resistance [24, 143, 158, 159]. Stabilized CK levels in transgenic arabidopsis plants lead to improved resistance to *Verticillium longispoum* [19, 160]. The interaction of ARABIDOPSIS RESPONSE REGULATOR 2 (ARR2), with TGA3 to promote plant defense response in an NPR 1 dependent manner is known [19, 160]. Likewise, rice resistance to *M. grisea* increases due to interaction between SA and CK in an OsNPR1 and WRKY45-dependent manner [14, 19]. Thus the role of cytokinins in the plant disease response cannot be over-emphasized [19, 152, 160, 161].

**Gibberellins (GA)**: Plant developmental processes including, seed development and seed germination, seedling growth, root proliferation, trichome initiation, determination of leaf size and shape, flower induction and development, pollination, fruit expansion etc. are mediated by gibberellins [162, 163]. Gibberellins are tetracyclic diterpenoid carboxylic acids; only a few of many known GAs, notably GA1 and GA4 act as plant hormones [164]. Gibberellins help plants to maintain their internal homeostasis by enabling control over their osmotic and water levels [162]. The mechanism of gibberellin action is relatively better understood in comparison to the other phytohormones. These work by bringing about the degradation of DELLA transcription factors via E3-ubiquitin-ligase [10, 165, 166]. In some plants loss of function mutations in DELLAs has been reported to improve the resistance of plants to biotic stress through SA dependent pathways, meaning thereby GAs, work in biotic as well as an abiotic stress response [10, 11, 167]. During seed germination and seedling establishment, exogenous GA reverses the inhibitory effect of the different stress factors and it also improves SA biosynthesis thereby, improving plant stress response [10]. GA antagonistically interacts with JA to mediate plant growth and defense response which involves direct interaction between DELLA and JAZ proteins [40, 168, 169]. Of the various JAZ proteins, osJAZ9 has been reported to be the key protein in mediating these interactions [40, 170]. The overall plant growth is the result of the fine balance between stress response and developmental process which is mediated through well-regulated JA/GA balance [34, 150, 171]. The "relief of repression" model explains this antagonistic interaction very well. It postulates that DELLAs and JAZ interact with each other leaving MYC2 free to mediated JA-dependent response under conditions of low GA while in presence of an adequate concentration of GA, DELLA is degraded by E3-ubiquitinylation mediated by GA leaving JAZ free to interact with MYC2 and attenuating the JA-mediated response [46]. Thus, JA/GA interact antagonistically [168]. The fact that JA promotes transcription of RGA3 (Repressor of GA1–3) and that MYC2 directly binds to their promoter further lends support to this model [40, 172].

**Brassinosteroids (BR):** First discovered from *Brassica napus*, BRs are polyhydroxy steroidal compounds about 70 different types of which have been isolated so far [46, 173]; only a few of them including brassinolide, 28-homobrassnolide and 24-epibrassinolide are actively engaged in the plant development [64, 174]. These are widely distributed in different plant organs including pollen, flower buds, vascular cambium, fruits, leaves, roots and shoots [64, 175]. These are also involved in modulating JA signaling and in the JA-dependent plant defense response. These affect many plant functions and alleviate the effects of hypoxia and unfavorable effects of various environmental stressors [46]. BRs are mainly seen as the hormones that alleviate abiotic stress, however, there are reports that these modulate the pattern-triggered immunity (PTI) (discussed in the next section) in *Arabidopsis* [176, 177].

## 4. Pathogen recognition reactions

A set of conserved pathogen proteins are important for plants to recognize the infection, these Microbe-Associated Molecular Patterns (MAMP), also called as Pathogen-Associated Molecular Patterns (PAMP), are recognized and bound by Pattern Recognition Receptors (PRRs) present in the host cell plasma membrane. This MAMP-PRR binding triggers an immune response called as MAMP Triggered Immunity (MTI) [24, 178, 179]. Microbes synthesize effectors which interfere with MTI and help pathogens evade recognition by the host immune system increasing their virulence and making plants susceptible to the pathogen and deregulating the host immunity, this process is known as Effector Triggered Susceptibility (ETS) [24, 180]. Bacteria acquire large repertoires of type III Effectors (T3E) and inject them through a syringe-like type III secretion system into their host plant. *Xanthomonas* sp. secretes Transcription activator-like (TAL) effectors, such as AvrBs3 secreted by *Xanthomonas axonopodis* pv. Vesicatoria, which after finding their way into the plant cell nucleus affects host gene expression [177, 181–183]. Auxin is a potential target for bacterial effectors. Effector proteins AvrBs3 1–5 have been reported to upregulate UPA1–5 [177, 184]. Induction of UPA20, a TAL target leads to cell hypertrophy indicating auxin accumulation [177, 185]. In *Arabidopsis* lines lacking the gene that recognizes T3E the bacterial effector AvrRpt2, a cysteine protease, triggers the auxin signaling pathway. Transgenic plants expressing AVrRpt2 accumulate higher auxin levels and constitutively express auxin signaling [177]. Thus, AVrRpt2 enhances bacterial virulence by affecting auxin signaling [177, 186]. An auxin signaling pathway is the preferred target of phytoplasmas [177, 187]. *Candidatus phytoplasma* asteris effector TENGU leads to dwarfism and abnormal organogenesis in reproductive parts leading to flower sterility. In transgenic *Arabidopsis* plants, many auxin-related genes including *Aux/IAA, SAUR, GH3* and *PIN* families were found to be downregulated indicating TENGU effector mediated disruption in the auxin signaling in the host plants [177, 188]. Similarly, *Ustilago maydis* hijacks the SA biosynthesis pathway in the maize plants to express its virulence [177]. Effector-triggered immunity (ETI) counters ETS in a gene-for-gene resistance mechanism [24, 189]. This leads to hypersensitivity response i.e., localized cell death in the infected region [24, 190, 191]. Hypersensitivity response leads to the activation of SAR [24]. The type of defense response depends upon the type of pathogens e.g., biotrophic pathogens are contained by programmed cell death which is mediated by SA [24]. On the other hand, the necrotrophs benefit from the cell death, the defense response, in this case, entails secretion of antibacterial/fungal compounds and accumulation of proteins that have antimicrobial properties such as defensins [24, 192–194]. JA-ET and SA, because of the inherent difference in the defense response these engage in, are antagonistic [24, 195, 196].

## 5. Hormone crosstalk in defense

Hormone crosstalk is the interaction of various plant hormones in a highly complex yet ordered manner [197, 198]. Crosstalk between various hormones is intensive in defense response [1, 5, 40, 45, 199]; it is mediated through regulatory proteins, hormone receptors, protein kinases, transcription factors etc. involved in hormone biosynthesis, degradation or signaling [5, 10, 40, 200, 201]. Hormonal cross-talk becomes increasingly important when plants are exposed to multiple pathogen stress simultaneously. Plants have to trade-off between defense and growth, therefore, the impact of individual hormones may not be as important as the overall interaction (positive or negative) among them. Numerous studies bring to light the flexible and coordinated interplay between growth and stress-related hormones especially JA and GA in regulating plant defense response [40, 150, 171, 202]; antagonistic interaction between SA and JA has been well researched [177, 203], the crosstalk between GA and SA was known only recently [10, 167]. Crosstalk enables pathogenesis-related genes affect response to abiotic stress [10, 204]. Crosstalk aids plants to gear up their defense system against various kinds of pathogens; however, all the aspects of this phenomenon in plant defense are not known.

## 6. Molecular Mechanism of hormone crosstalk

Phosphorylation Cascade is a common second messenger which integrates various hormone responses. JAZ and DELLA proteins mediate the antagonistic interactions between JA and GA [198, 205, 206]. A great deal of information exists on how DELLAs interact with JAZ proteins [168, 169]. GA response in *Arabidopsis* is suppressed by direct interaction of its transcription-factors-like PIF (phytochrome interacting factors) with the DELLAs including GIBBERELLIC ACID INSENSITIVE (GAI), REPRESSOR OF GA (RGA), RGA-like1 (RGL1), RGL2 and RGL3 thus, DELLAs act as negative regulators of the GA response [198, 207]. DELLAs binding with JAZ1 leaves MYC2 free to initiate JA signal response thereby enabling JA-responsive transcription [19, 168]. In a contrary situation, higher GA levels attenuate JA signaling by degrading DELLAs thereby allowing interaction of MYC2 with JAZ1 [40, 198, 208]. GA-related transcription factors like PIF3 are repressed at higher JA levels as JA stabilizes DELLA proteins through JAZ degradation [169, 198]. JA/GA antagonism in rice is mediated by an interaction between OsJAZ9 and DELLA proteins, namely, SLENDER RICE 1 (SLR1) [170, 198]. DELLAs also repress SA biosynthesis as well signaling affecting the balance between JA and SA [10, 167]. JA leads to selection of defense overgrowth in the events of pathogen attack by interfering with GA-mediated degradation of DELLAs [19, 169, 209]. In DELLA quadruple mutants (mutant lacking GAI, RGA, RGL1 and RGL2 proteins) expression of PR1 and PR2 is increased which makes them more resistant to hemi-biotrophs, however, delayed induction of PDF1.2 a JA/ET dependent gene marker in such mutants leaves them susceptible to necrotrophs [19, 167]. By way of controlling DELLAs GAs indirectly control the SA/JA balance. The molecular mechanism of antagonistic interaction of SA with ET and JA pathways is largely unknown [1]. NPR1, however, is at the core of most of these antagonistic interactions. A WRKY70 transcription factor is another key player in the hormone crosstalk. WRKY33 is a positive regulator of JA-dependent genes but a repressor of SA pathway, therefore, wrky33 mutants show upregulated expression of several SA-regulated genes including SID2/ICS1, EDS5/SID1, PAD4, EDS1, NIMIN1, PR1, PR2, PR3. SA induction leads to down-regulated JA signaling and increased susceptibility of these mutants to necrotrophic fungi [177, 210]. When over-expressed it leads to constitutive expression of SA-responsive PR genes and repression of JA responsive PDG1.2 gene [19, 211]. Likewise, *Arabidopsis* mpk4 (MAP kinase 4) knockout mutants exhibit constitutive SAR, higher expression of PR genes but an impaired expression of JA-responsive *PDF1.2* and *THI2.1* genes [19, 196]. Synergistic interactions between SA and JA have also been reported especially at their lower concentrations and when both the defense responses are triggered together [19, 196, 211]. MED16 which positively regulates SA-induced defense response negatively regulates JA/ET signaling pathway [177, 212]. Some strains of *P. syringae* produce phytotoxin coronatine (COR), a mimic of the JA-Ile and this suppresses SA signaling [177, 213, 214]. This is the reason for the lower virulence of the strains of *P. syringae* that have impaired production of COR on wild *Arabidopsis* plants but not on SA deficient plants [177, 215]. JA and ET cooperate in comparison with the JA and SA where interactions are mostly antagonistic, e.g., JA and ET both stabilize EIN3 thereby leading to the defense of roots against necrotrophic pathogens [19, 56]. Both JA and ET activate the expression of ERF1 which in turn activates PR genes [19, 216]. However, in cases of herbivore and insect attacks the two pathways may interact antagonistically e.g., JA–activated MYC2 interacts with ET-stabilized EIN3 and represses its downstream activity. In turn, EIN3 represses MYC2 thereby repressing JA-mediated defense response against herbivores [19, 217].

## 7. Concluding remarks and future perspectives

Allocation of resources and energy to defense in absence of threat would constrain growth and developmental processes [177, 218–220]. Therefore, a hormone-based defense mechanism in plants evolved to prevent loss of resources in absence of stress [177, 220] slowing down the potential adaptation of putative attackers to the biochemical defense system of plants [177, 220]. During priming plants subjected to pathogen attack respond more strongly to subsequent pathogen attacks, resources here are not committed until the threat returns making priming a relatively cost-effective defense strategy [177, 221]. Moreover, the primed plants treated with a low, non-effective concentration of defense hormones also respond better to the pathogen attack than the non-primed ones [177, 221]. Priming has parallels with the trans-generational defense in plants, such as SA-dependent SAR and JA-dependent inherited defense as trans-generational priming has been described in some plants [80, 177, 214]. Epigenetically inherited changes can strongly affect the defense response including priming in plants [80, 177, 222].

Our understanding of plant defense response has considerably improved in the past few years due to modified and transgenic plants species [177]. Transgenic plants constitutively expressing some hormones have been reported to show improved resistance to pathogens [177, 223, 224]. But, such an "effective" resistance response is also known to incur the costs paid in terms of altered development e.g. dwarfism, development of spontaneous lesions in different organs, accelerated pace of senescence, delayed flowering, sterility and lower seed output [177, 223–225].

Dissecting hormone response specifically in the event of a pathogen attack is complicated by the complex regulatory pathways interconnecting at several different levels. In nature, a plant has to deal with both abiotic and biotic stresses therefore, its response to the environment, in general, has to be concerted and balanced. Ideally, a plant resistant to biotic or abiotic stress should not be hampered in terms of its growth, development and overall productivity. It is a generally conceded fact that the traditional methods of crop improvement have reached their peak and are now leveling off. Thus, molecular and genetic engineering methods provide reliable alternative means of crop improvement. Phytohormone engineering is seen as a new opportunity to maintain susceptible crop production, especially in the climate change scenario. Elucidating the path of signal transduction in stress response is an important step in manipulating the role of phytohormones in stress response. In the future plant defense response mediated by hormones should be studied under field conditions with model crop plants so that a better picture of the effectiveness of the hormone-mediated disease control, associated trade-offs in growth and development parameters, and impact on the performance of the plants are brought to light. A clear understanding of the hormone homeostasis at the molecular level is required to manipulate it and use it as a tool for effective defense against crop pathogens.

## Author details

Nazima Rasool
University of Kashmir, Kargil Campus, India

*Address all correspondence to: rasoolnazima@gmail.com

## References

[1] Ning L, Xiao H, Feng D, Yuan D, Huang L. Signaling crosstalk between salicylic acid and ethylene/jasmonate in plant defense: Do we do we understand what they are whispering? International Journal of Molecular Sciences. 2019;**20**(3):671. DOI: 10.3390/ijms20030671

[2] Madani B, Mirshekari A, Imahori Y. Chapter 19 - physiological responses to stress. In: Yahia EM, editor. Postharvest Physiology and Biochemistry of Fruits and Vegetables. Sawton: Woodhead Publishing; 2019. pp. 405-423

[3] Großkinsky DK, van der Graaff E, Roitsch T. Regulation of abiotic and biotic stress responses by plant hormones. Plant Pathogen Resistance Biotechnology. 2016:131

[4] Liu Y, Du M, Deng L, Shen J, Fang M, Chen Q, et al. MYC2 regulates the termination of jasmonate signaling via an autoregulatory negative feedback loop. Plant Cell. 2019;**31**:106-127

[5] Yang J, Duan G, Li C, Liu L, Han G, Zhang Y, et al. The crosstalks between jasmonic acid and other plant hormone signaling highlight the involvement of jasmonic acid as a core component in plant response to biotic and abiotic stresses. Frontiers in Plant Science. 2019;**10**:1349

[6] Wang J, Song L, Gong X, Xu J, Li M. Functions of jasmonic acid in plant regulation and response to abiotic stress. International Journal of Molecular Sciences. 2020;**21**(4):1446

[7] Rajjou L, Belghazi M, Huguet R, Robin C, Moreau A, Job C, et al. Proteomic investigation of the effect of salicylic acid on Arabidopsis seed germination and establishment of early defense mechanisms. Plant Physiology. 2006;**141**:910-923

[8] Gao Q-M, Zhu S, Kachroo P, Kachroo A. Signal regulators of systemic acquired resistance Front. Plant Science. 2015;**6**(228):1-12

[9] Li J, Brader G, Kariola T, Tapio Palva E. WRKY70 modulates the selection of signaling pathways in plant defense. The Plant Journal. 2006;**46**:477-491

[10] Alonso-Ramırez A, Rodrıguez D, Reyes D, Jimenez JA, Nicolas G, Lopez-Climent M, Gomez-Cadenas A, and Nicolas C. Evidence for a Role of Gibberellins in Salicylic Acid-Modulated Early Plant Responses to Abiotic Stress in Arabidopsis Seeds Plant Physiology. 2009;**150**:1335-1344

[11] Robert-Seilaniantz A, Navarro L, Bari R, Jones JD. Pathological hormone imbalances. Current Opinion in Plant Biology. 2007;**10**:372-379

[12] Shi H, Chen L, Ye T, Liu X, Ding K, Chan Z. Modulation of auxin content in Arabidopsis confers improved drought stress resistance. Plant Physiology and Biochemistry. 2014;**82**:209-217

[13] Ghanashyam C, Jain M. Role of auxin-responsive genes in biotic stress responses. Plant Signaling & Behavior. 2009;**4**(9):846-848

[14] Jiang CJ, Shimono M, Sugano S, Kojima M, Liu X, Inoue H, et al. Cytokinins act synergistically with salicylic acid to activate defense gene expression in rice. Molecular Plant-Microbe Interactions. 2013;**26**(3):287-296

[15] Dolgikh VA, Pukhovaya EM, Zemlyanskaya EV. Shaping Ethylene Response: The Role of EIN3/EIL1 Transcription Factors. Front Plant Science. 2019;**26** 1030

[16] Ecker JR. The ethylene signal transduction pathway in plants. Science. 1995;**268**:667-675

[17] Bari R, Jones JD. Role of plant hormones in plant defense responses. Plant Molecular Biology. 2009;**69**(4):473-488

[18] Gamalero E, Glick BR. Ethylene and abiotic stress tolerance in plants. In: Environmental Adaptations and Stress Tolerance of Plants in the Era of Climate Change. New York: Springer; 2012. pp. 395-412.

[19] Verma, Vivek, Ravindran, P., and Kumar P. P., Plant hormone-mediated regulation of stress responses BMC Plant Biology. 2016;**16**:86. DOI: 10.1186/s12870-016-0771-y

[20] Morgan PW, Drew MC. Ethylene and plant responses to stress. Physiologia Plantarum. 1997;**100**: 620-630

[21] Wang D, Pei K, Fu Y, Sun Z, Li S, Liu H, et al. Genome-wide analysis of the auxin response factors (ARF) gene family in rice (Oryza sativa). Gene. 2013 2007;**394**:13-24

[22] Wang FF, Cui XK, Sun Y, Dong CH. Ethylene signaling and regulation in plant growth and stress responses. Plant Cell Reports. 2013;**32**:1099-1109

[23] Zwack PJ, Rashotte AM. Interactions between cytokinin signaling and abiotic stress responses. Journal of Experimental Botany. 2015;**66**: 4863-4871

[24] O'Brien JA, Benková E. Cytokinincross-talking during biotic and abiotic stress responses. Front Plant Sci. 2013;**4**:451

[25] Bleecker AB, Estelle MA, Somerville C, Kende H. Insensitivity to ethylene conferred by a dominant mutation in *Arabidopsis thaliana*. Science. 1988;**241**:1086-1089

[26] Chang C, Kwok SF, Bleecker AB, Meyerowitz EM. Arabidopsis ethylene-response gene ETR1: Similarity of product of two-component regulators. Science. 1993;**262**:539-544

[27] Hua J, Sakai H, Nourizadeh S, Chen QG, Bleecker AB, Ecker JR, et al. EIN4 and ERS2 are members of the putative ethylene receptor gene family in Arabidopsis. Plant Cell. 1998;**10**:1321-1332

[28] Sakai H, Hua J, Chen QG, Chang C, Medrano LJ, Bleecker AB, et al. ETR2 is an ETR1-like gene involved in ethylene signaling in Arabidopsis. Proceedings. National Academy of Sciences. United States of America 1998;**95**:5812-5817. DOI: 10.1073/pnas.95.10. 5812

[29] Geraats BPJ, Bakker PAHM, Lawrence CB, Achuo EA, Höfte M, van Loon LC. Ethylene-insensitive tobacco shows differentially altered susceptibility to different pathogens. Phytopathology. 2003;**93**:813-821. DOI: 10.1094/PHYTO.2003.93.7.813

[30] Challa G, Jain M. Role of auxin-responsive genes in biotic stress responses. Plant Signaling & Behavior. Landes: Bioscience; 2009;**4**(9):846-848

[31] Cho YH, Yoo SD. Novel connections and gaps in ethylene signaling from the ER membrane to the nucleus. Frontiers in Plant Science. 2015;**5**:733. DOI: 10.3389/fpls.2014.00733

[32] Guo H, Ecker JR. The ethylene signaling pathway: new insights. Current Opinion in Plant Biology. 2004;**7**:40-49. DOI: 10.1016/j.pbi.2003.11.011

[33] Chang C. Q&A: how do plants respond to ethylene and what is its importance? BMC Biology. 2016;**14**:7

[34] Hu Y, Vandenbussche F, Van Der Straeten D. Regulation of seedling growth by ethylene and the ethylene–auxin crosstalk. Planta. 2017;**245**:467-489. DOI: 10.1007/s00425-017-2651-6

[35] Solano R, Stepanova A, Chao Q, Ecker JR. Nuclear events in ethylene signaling: a transcriptional Cascade mediated by ETHYLENEINSENSITIVE3 and ETHYLENE-RESPONSE-FACTOR1. Genes & Development. 1998;**12**(23):3703-3714

[36] Ju C, Yoon GM, Shemansky JM, Lin DY, Ying ZI, Chang J, et al. CTR1 phosphorylates the Central regulator EIN2 to control ethylene hormone signaling from the ER membrane to the nucleus in Arabidopsis. Proceedings of the National Academy of Sciences of The United States of America. 2012;**109**(47):19486-19491

[37] Li W, Ma M, Feng Y, Li H, Wang Y, Ma Y, et al. EIN2-directed translational regulation of ethylene signaling in arabidopsis. Cell. 2015;**163**:670-683. DOI: 10.1016/j.cell.2015.09.037

[38] Merchante C, Brumos J, Yun J, Hu Q, Spencer KR, Enríquez P, et al. Gene-specific translation regulation mediated by the hormone-signaling molecule EIN2. Cell. 2015;**163**:684-697

[39] An F, Zhao Q, Ji Y, Li W, Jiang Z, Yu X, et al. Ethylene-induced stabilization of ETHYLENE INSENSITIVE3 and EIN3-LIKE1 is mediated by proteasomal degradation of EIN3 binding F-box 1 and 2 that requires EIN2 in Arabidopsis. Plant Cell. 2010;**22**:2384-2401

[40] Jang G, Yoon Y, Choi YD. Crosstalk with jasmonic acid integrates multiple responses in plant development International Journal of Molecular Sciences. 2020;**21**:305

[41] Fahad SL, Nie Y, Chen C, Wu D, Xiong S, Saud L, et al. Crop plant hormones and environmental stress. Sustain. Agric. Rev. 2015;**15**:371-400

[42] Pauwels L, Inze D, Goossens A. Jasmonate-inducible gene: what does it mean? Trends in Plant Science. 2009;**14**:87-91

[43] Seo JS, Joo J, Kim MJ, Kim YK, Nahm BH, Song SI, et al. OsbHLH148, a basic helix-loop-helix protein, interacts with OsJAZ proteins in a jasmonate signaling pathway leading to drought tolerance in rice. The Plant Journal. 2011;**65**:907-921

[44] Shabir W, Vinay Kumar H, Shriram V, Sah SK. Phytohormones and their metabolic engineering for abiotic stress tolerance in crop plants. CAAS: Crop Science Society of China and Institute of Crop Science; 2016. pp. 2214-5141

[45] de Ollas C, Dodd IC. Physiological impacts of ABA–JA interactions under water-limitation. Plant Molecular Biology. 2016;**91**(6):641-650

[46] Miransari M. Role of phytohormone signaling during stress. In: Ahmad P, Prasad MNV, editors. Environmental Adaptations and Stress Tolerance of Plants in the Era of Climate Change. 1st ed. New York: Springer; 2012

[47] Schaller A, Stintzi A. Enzymes in jasmonate biosynthesis – Structure, function, regulation. Phytochemistry. 2009;**70**:1532-1538

[48] Devoto A, Nieto-Rostro M, Xie D, Ellis C, Harmston R, Patrick E, et al. COI1 links jasmonate signaling and fertility to the SCF ubiquitin-ligase complex in Arabidopsis. The Plant Journal. 2002;**32**:457-466

[49] Xu L, Liu F, Lechner E, Genschik P, Crosby WL, Ma H, et al. The SCFCOI1 ubiquitin-ligase complexes are required for jasmonate response in Arabidopsis. Plant Cell. 2002;**14**:1919-1935

[50] Eulgem T, Somssich IE. Networks of WRKY transcription factors in defense signaling. Current Opinion in Plant Biology. 2007;**10**(4):366-371

[51] Zhang F, Yao J, Ke J, Zhang L, Lam VQ, Xin X-F, et al. Structural basis

of JAZ repression of MYC transcription factors in jasmonate signalling. Nature. 2015;**525**:269-273

[52] Berrocal-Lobo M, Molina A, Solano R. Constitutive expression of ETHYLENERESPONSE-FACTOR1 in Arabidopsis confers resistance to several necrotrophic fungi. The Plant Journal. 2002;**29**(1):23-32

[53] Brown RL, Kazan K, McGrath KC, Maclean DJ, Manners JM. A role for the GCC-box in jasmonate-mediated activation of the PDF1.2 gene of Arabidopsis. Plant Physiology. 2003;**132**(2):1020-1032

[54] Moffat CS, Ingle RA, Wathugala DL, Saunders NJ, Knight H, Knight MR. ERF5 and ERF6 play redundant roles as positive regulators of JA/Etmediated defense against Botrytis cinerea in Arabidopsis. PLoS One. 2012;**7**(4):e35995

[55] Pré M, Atallah M, Champion A, De Vos M, Pieterse CMJ, Memelink J. The AP2/ERF domain transcription factor ORA59 integrates jasmonic acid and ethylene signals in plant defense. Plant Physiology. 2008;**147**:1347-1357

[56] Zhu Z, An F, Feng Y, Li P, Xue L, Mu A, et al. Depression of ethylene-stabilized transcription factors (EIN3/EIL1) mediates jasmonate and ethylene signaling synergy in Arabidopsis. Proc. Natl. Acad. Sci. USA. 2011;**108**: 12539-12544

[57] AAdie BAT, Pérez-Pérez J, Pérez-Pérez MM, Godoy M, Sánchez-Serrano J-J, Schmelz EA, et al. ABA is an essential signal for plant resistance to pathogens affecting JA biosynthesis and the activation of defenses in Arabidopsis. Plant Cell. 2007;**19**:1665-1681

[58] Ferrari S, Galletti R, Denoux C, DeLorenzo G, Ausubel FM, Ewdney J. Resistance to Botrytis cinerea induced in Arabidopsis by elicitor is independent of salicylic acid, ethylene, or jasmonate signaling but requires PHYTOALEXINDEFICIENT3. Plant Physiology. 2007;**144**:367-379

[59] vanWees SCM, Chang H-S, Zhu T, Glazebrook J. Characterization of the early response of Arabidopsis to Alternaria brassicicola infection using expression profiling. Plant Physiology. 2003;**132**:606-617. DOI: 10.1104/pp.103.022186

[60] Ye M, Luo SM, Xie JF, Li YF, Xu T, Liu Y, et al. Silencing COI1 in rice increases susceptibility to chewing insects and impairs inducible defense. PLoS One. 2012;7:e36214

[61] Chung HS, Koo AJK, Gao X, Jayanty S, Thines B, Jones AD, et al. Regulation and function of Arabidopsis JASMONATEZIM-domain genes in response to wounding and herbivory. Plant Physiology. 2008;**146**:952-964

[62] Jung C, Zhao P, Seo JS, Mitsuda N, Deng S, Chua NH. Plant U-BOX protein10 regulates MYC2 stability in arabidopsis. The Plant Cell. 2015;**27**(7):2016-2031

[63] Miura K, Tada Y. Regulation of water, salinity, and cold stress responses by salicylic acid. Frontiers in Plant Science. 2014;**5**:4

[64] Wani SH, Kumar V, Shriram V, Sah SK. Phytohormones and their metabolic engineering for abiotic stress tolerance in crop plants. The Crop Journal. 2016;**4**(3):162-176

[65] Jumali SS, Said IM, Ismail I, Zainal Z. Genes induced by high concentration of salicylic acid in Mitragyna speciosa. Australian Journal of Crop Science. 2011;**5**:296-303

[66] Horvath E, Szalai G, Janda T. Induction of abiotic stress tolerance by salicylic acid signaling. Journal of Plant Growth Regulation. 2007;**26**:290-300

[67] van Loon LC, Rep M, Pieterse CM. Significance of inducible defense related proteins in infected plants. Annual Review of Phytopathology. 2006;**44**:135-162

[68] Nawrath C, Métraux JP. Salicylic acid induction- deficient mutants of Arabidopsis express PR-2 and PR-5 and accumulate high levels of camalexin after pathogen inoculation. Plant Cell. 1999;**11**:1393-1404

[69] Delaney TP, Uknes S, Vernooij B, Friedrich L, Weymann K, Negrotto D, et al. A Central role of salicylic acid in plant disease resistance. Science. 1994;**266**:1247-1250

[70] Gaffney T, Friedrich L, Vernooij B, Negrotto D, Nye G, Uknes S, et al. Requirement of salicylic acid for the induction of systemic acquired resistance. Science. 1993;**261**:754-756

[71] Chen H, Xue L, Chintamanani S, Germain H, Lin H, Cui H, et al. Ethylene insensitive3 and ethylene insensitive3-like1 repress salicylic acid induction deficient2 expression to negatively regulate plant innate immunity in Arabidopsis. Plant Cell. 2009;**21**:2527-2540. DOI: 10.1105/tpc.108.065193

[72] Dong X. NPR1, all things considered. Current Opinion in Plant Biology. 2004;**7**:547-552

[73] Mou Z, Fan W, Dong X. Inducers of plant systemic acquired resistance regulate NPR1 function through redox changes. Cell. 2003;**113**:935-944

[74] Tada Y, Spoel SH, Pajerowska-Mukhtar K, Mou Z, Song J, Wang C, et al. Plant immunity requires conformational changes of NPR1 via S-nitrosylation and thioredoxins. Science. 2008;**321**:952-956

[75] Loake G, Grant M. Salicylic acid in plant defence—the players and protagonists. Current Opinion in Plant Biology. 2007;**10**(5):466-472

[76] Wang Y, Li L, Ye T, Zhao S, Liu Z, Feng YQ, et al. Cytokinin antagonizes ABA suppression to seed germination of Arabidopsis by downregulating ABI5 expression. The Plant Journal. 2011a;**68**:249-261

[77] Zhang K, Halitschke R, Yin C, Liu CJ, Gan SS. Salicylic acid 3-hydroxylase regulates Arabidopsis leaf longevity by mediating salicylic acid catabolism. Proceedings. National Academy of Sciences. United States of America. 2013;**110**:14807-14812

[78] Spoel SH, Mou Z, Tada Y, Spivey NW, Genschik P, Dong X. Proteasome-mediated turnover of the transcription coactivator NPR1 plays dual role in regulating plant immunity. Cell. 2009;**137**:860-872

[79] Fu ZQ, Yan S, Saleh A, Wang W, Ruble J, Oka N, et al. NPR3 and NPR4 are receptors for the immune signal salicylic acid in plants. Nature. 2012;**486**:228-232. DOI: 10.1038/nature11162

[80] Luna E, Bruce TJ, Roberts MR, Flors V, Ton J. Next-generation systemic acquired resistance. Plant Physiology. 2012;**158**:844-853. DOI: 10.1104/pp.111.187468

[81] Yuan Y, Zhong S, Li Q, Zhu Z, Lou Y, Wang L, et al. Functional analysis of rice NPR1-like genes reveals that OsNPR1/NH1 is the rice orthologue conferring disease resistance with enhanced herbivore susceptibility. Plant Biotechnology Journal. 2007;**5**:313-324

[82] Kazan K. Auxin and the integration of environmental signals into plant root development. Annals of Botany. 2013;**112**:655-1665

[83] Bouzroud S, Gouiaa S, Hu N, Bernadac A, Mila I, Bendaou N, et al. Auxin Response Factors (ARFs) are potential mediators of auxin action in tomato response to biotic and abiotic

stress (*Solanum lycopersicum*). PLoS One. 2018;**13**(2):e0193517

[84] Causier B, Lloyd J, Stevens L, Davies B. TOPLESS co-repressor interactions and their evolutionary conservation in plants. Plant Signaling & Behavior. 2012;**7**:325-328

[85] Jain M, Khurana JP. Transcript profiling reveals diverse roles of auxin-responsive genes during reproductive development and abiotic stress in rice. The FEBS Journal. 2009;**276**:3148-3162

[86] Szemenyei H, Hannon M, Long JA. TOPLESS mediates auxin-dependent transcriptional repression during Arabidopsis embryogenesis. Science. 2008;**319**:1384-1386

[87] Zouine M, Fu Y, Chateigner-Boutin A-L, Mila I, Frasse P, Wang H, et al. Characterization of the tomato ARF gene family uncovers a multi-levels post-transcriptional regulation including alternative splicing. PLoS One. 2014;**9**:e84203

[88] Liscum E, Reed J. Genetics of Aux/IAA and ARF action in plant growth and development. Plant Molecular Biology. 2002;**49**:387-400.

[89] Xing H, Pudake RN, Guo G, Xing G, Hu Z, Zhang Y, et al. Genome-wide identification and expression profiling of auxin response factor (ARF) gene family in maize. BMC Genomics. 2011;**12**:178

[90] Wang D, Pei K, Fu Y, Sun Z, Li S, Liu H, et al. Genome-wide analysis of the auxin response factors (ARF) gene family in rice (Oryza sativa). Gene. 2007;**394**:13-24

[91] Kalluri UC, DiFazio SP, Brunner AM, Tuskan GA. Genome-wide analysis of Aux/IAA and ARF gene families in Populus trichocarpa. BMC Plant Biology. 2007;**7**:59

[92] TGG: Tomato Genome Consortium. The tomato genome sequence provides insights into fleshy fruit evolution. Nature 2012;**485**:635. DOI: 10.1038/nature11119

[93] Wu J, Wang F, Cheng L, Kong F, Peng Z, Liu S, et al. Identification, isolation and expression analysis of auxin response factor (ARF) genes in Solanum lycopersicum. Plant Cell Reports. 2011;**30**:2059

[94] Mun J-H, Yu H-J, Shin JY, Oh M, Hwang H-J, Chung H. Auxin response factor gene family in Brassica rapa: Genomic organization, divergence, expression, and evolution. Molecular Genetics and Genomics. 2012;**287**: 765-784

[95] Van Ha C, Le DT, Nishiyama R, Watanabe Y, Sulieman S, Tran UT, et al. The auxin response factor transcription factor family in soybean: genome-wide identification and expression analyses during development and water stress. DNA Res Int J Rapid Publ Rep Genes Genomes. 2013;**20**:511-524

[96] Hu W, Zuo J, Hou X, Yan Y, Wei Y, Liu J, et al. The auxin response factor gene family in banana: Genome-wide identification and expression analyses during development, ripening, and abiotic stress. Front Plant Science. 2015;**6**(742):1-16

[97] Dowd C, Wilson IW, McFadden H. Gene expression profile changes in cotton root and hypocotyl tissues in response to infection with *Fusarium oxysporum* f.sp. *vasinfectum*. Molecular Plant-Microbe Interactions. 2004;**17**: 654-667

[98] Llorente F, Muskett P, Sánchez-Vallet A, López G, Ramos B, Sánchez-Rodríguez C, et al. Repression of the auxin response pathway increases Arabidopsis susceptibility to necrotrophic fungi. Molecular Plant. 2008;**1**:496-509

[99] Bishopp A, Help H, El-Showk S, Weijers D, Scheres B, Friml J, et al. A mutually inhibitory interaction between auxin and cytokinin specifies vascular pattern in roots. Current Biology. 2011;**21**:927-932

[100] Vanstraelen M, Benková E. Hormonal interactions in the regulation of plant development. Annual Review of Cell and Developmental Biology. 2012;**28**:463-487

[101] Ding X, Cao Y, Huang L, Zhao J, Xu C, Li X, et al. Activation of the indole-3- acetic acid amidosynthetase GH3-8 suppresse expansin expression and promote ssalicylate-and jasmonate-independent basal immunity in rice. Plant Cell. 2008;**20**:228-240

[102] Chen Z, Agnew JL, Cohen JD, He P, Shan L, Sheen J, et al. Pseudomonas syringae type III effector AvrRpt2 alters Arabidopsis thaliana auxin physiology. Proceedings. National Academy of Sciences. United States of America 2007;**104**:20131-20136

[103] Naseem M, Philippi N, Hussain A, Wangorsch G, Ahmed N, Dandekar T. Integrated systems view on networking by hormones in Arabidopsis immunity reveals multiple crosstalk for cytokinin. Plant Cell. 2012;**24**:1793-1814

[104] Naseem M, Dandekar T. The role of auxin-cytokinin antagonism in plant-pathogen interactions. PLoS Pathogens 2012;**8**:e1003026. DOI: 10.1371/journal.ppat.1003026

[105] Liu J, Wang X-J. An integrative analysis of the effects of auxin on jasmonic acid biosynthesis in *Arabidopsis thaliana*. Journal of Integrative Plant Biology. 2006;**48**: 99-103

[106] Zhang Z, Li Q, Li Z, Staswick PE, Wang M, Zhu Y, et al. Dual regulation role of *GH3.5* in salicylic acid and auxin signaling during Arabidopsis-*Pseudomonas syringae* interaction. Plant Physiology. 2007;**145**:450-464

[107] Nambara E. Encyclopedia of Applied Plant Sciences (Second Edition). Plant Physiology and Development. Academic Press, Elsevier; 2017

[108] Cutler SR, Rodriguez PL, Finkelstein RR, Abrams SR. Abscisic acid: Emergence of a core signaling network. Annual Review of Plant Biology. 2010;**61**:651-679

[109] Danquah A, de Zelicourt A, Colcombet J, Hirt H. The role of ABA and MAPK signaling pathways in plant abiotic stress responses. Biotechnology Advances. 2014;**32**:40-52

[110] Nemhauser JL, Hong F, Chory J. Different plant hormones regulate similar processes through largely non overlapping transcriptional responses. Cell. 2006;**126**:467-475

[111] Chen K, Li GJ, Bressan RA, Song CP, Zhu JK, Zhao Y. Abscisic acid dynamics, signaling, and functions in plants. Journal of Integrative Plant Biology. Jan 2020;**62**(1):25-54

[112] Finkelstein R. Abscisic Acid Synthesis and Response American Society of Plant Biollogists. 2013;**11**: e0166

[113] Asselbergh B, De Vleesschauwer D, Hofte M. Global switches and fine-tuning-ABA mod- ulates plant pathogen defense. Molecular Plant-Microbe Interactions. 2008;**21**:709-719

[114] Feng DX, Tasset C, Hanemian M, Barlet X, Hu J, Trémousaygue D, et al. Biological control of bacterial wilt in Arabidopsis thaliana involves abscisic acid signalling. The New Phytologist. 2012;**194**:1035-1045

[115] Mauch-Mani B, Mauch F. The role of abscisic acid in plant– pathogen

interactions. Curr.Opin. Plant. Biol. 2005;**8**:409-414

[116] Ton J, Flors V, Mauch-Mani B. The multifaceted role of ABA in disease resistance. Trends in Plant Science. 2009;**14**:310-317

[117] Audenaert K, De Meyer GB, Hofte MM. Abscisic acid determines basal susceptibility of tomato to Botrytis cinerea and sup- presses salicylic acid-dependent signaling mechanisms. Plant Physiology 2002;**128**:491-501

[118] de Torres-Zabala M, Truman W, Bennett MH, Lafforgue G, Mansfield JW, Rodriguez Egea P, et al. Pseudomonas syringae pv. tomato hijacks the Arabidopsis abscisic acid signaling pathway to cause disease. The EMBO Journal. 2007;**26**:1434-1443

[119] de TorresZabala M, Bennett MH, Truman WH, andGrant, M.R. Antagonism between salicylic and abscisic acid reflects early host–pathogen conflict and moulds plant defence responses. The Plant Journal. 2009;**59**:375-386

[120] Garcia-Andrade J, Ramirez V, Flors V, Vera P. Arabidopsis ocp3 mutant reveals a mechanism linking ABA and JA to pathogen- induced callose deposition. The Plant Journal. 2011;**67**:783-794

[121] Mohr PG, Cahill D. Abscisic acid influences the susceptibility of Arabidopsis thaliana to Pseudomonas syringae pv. tomato and Peronospora parasitica. Functional Plant Biology 2003;**30**:461-469

[122] Sánchez-Vallet A, Lopez G, Ramos B, Delgado-Cerezo M, Riviere MP, Llorente F, et al. Disruption of abscisic acid signalling constitutively activates Arabidopsis resistance to the necrotrophic fungus Plectosphaerella cucumerina. Plant Physiology. 2012;**160**:2109-2124

[123] Yasuda M, Ishikawa A, Jikumaru Y, Seki M, Umezawa T, Asami T, et al. Antagonistic interaction between systemic acquired resistance and the abscisic acid-mediated abiotic stress response in Arabidopsis. Plant Cell. 2008;**20**:1678-1692

[124] Mang HG, Qian W, Zhu Y, Qian J, Kang HG, Klessig DF, et al. Abscisic acid deficiency antagonizes high-temperature inhibition of disease resistance through enhancing nuclear accumulation of resistance proteins SNC1 and RPS4 in Arabidopsis. Plant Cell. 2012;**24**:1271-1284

[125] Flors V, Ton J, van Doorn R, Jakab G, Garcia-Agustin P, Mauch-Mani B. Interplay between JA, SA and ABA signaling during basal and induced resistance against *Pseudomonas syringae* and *Alternaria brassicicola*. The Plant Journal 2008;**54**:81-92

[126] Hernandez-Blanco C, Feng DX, Hu J, Sanchez-Vallet A, Deslandes L, Llorente F, et al. Impairment of cellulose synthases required for Arabidopsis secondary cell wall formation enhances disease resistance. Plant Cell. 2007;**19**:890-903

[127] Kieber JJ, Schaller GE. Cytokinin signaling in plant development. Development. 2018;**145**:149344

[128] Schaller GE, Street IH, Kieber JJ. Cytokinin and the cell cycle. Current Opinion in Plant Biology. 2014;**21**:7-15

[129] Bielach A, Podlešáková K, Marhavý P, Duclercq J, Cuesta C, Müller B, Grunewald W, Tarkowski P, Benková E. Spatiotemporal regulation of lateral root organogenesis in Arabidopsis by cytokinin. The Plant Cell. 2012;**24**:3967-3981

[130] Gan S, Amasino R. Cytokinins in plant senescence: from spray and pray to clone and play. BioEssays. 1996;**18**:557-565

[131] Laplaze L, Benkova E, Casimiro I, et al. Cytokinins act directly on lateral root founder cells to inhibit root initiation. The Plant Cell; 2007;**19**:3889-3900.

[132] Zwack PJ, Rashotte AM. Cytokinin inhibition of leaf senescence. Plant Signaling and Behavior. 2013;**8**:24737e

[133] Bishopp A, Benková E, Helariutta Y. Sending mixed messages: auxin-cytokinin crosstalk in roots. Current Opinion in Plant Biology. 2011b;**14**:10-16

[134] Mähönen AP, Bonke M, Kauppinen L, Riikonon M, Benfey P, Helariutta Y. A novel two-component hybrid molecule regulates vascular morphogenesis of the Arabidopsis root. Genes and Development. 2000;**14**: 2938-2943

[135] Chiang YH, Zubo YO, Tapken W, Kim HJ, Lavanway AM, Howard L, et al. The GATA transcription factors GNC and CGA1 positively regulate chloroplast development, growth, and division in Arabidopsis. Plant Physiology. 2012;**160**:332-348

[136] Chory J, Reinecke D, Sim S, Washburn T, Brenner M. A role for cytokinins in de-etiolation in Arabidopsis. Plant Physiology. 1994;**104**:339-347

[137] Efroni I, Han SK, Kim Hye J, Wu MF, Steiner E, Birnbaum KD, et al. Regulation of leaf maturation by chromatin-mediated modulation of cytokinin responses. Developmental Cell. 2013;**24**:438-445

[138] Miyawaki K, Tarkowski P, Matsumoto-Kitano M, Kato T, Sato S, Tarkowska D, et al. Roles of Arabidopsis ATP/ADP isopentenyltransferases and tRNA isopentenyltransferases in cytokinin biosynthesis. Proc. Natl. Acad. Sci. USA. 2006;**103**:16598-16603

[139] Gajdošová S, Spíchal L, Kamínek M, Hoyerová K, Novák O, Dobrev PI, et al. Distribution, biological activities, metabolism, and the conceivable function of cis-zeatin-type cytokinins in plants. Journal of Experimental Botany. 2011;**62**: 2827-2840

[140] Schäfer M, Brütting C, Meza-Canales ID, Großkinsky DK, Vankova R, Baldwin IT, et al. The role of cis-zeatin-type cytokinins in plant growth regulation and mediating responses to environmental interactions. Journal of Experimental Botany. 2015;**66**:4873-4884

[141] Masuta C, Tanaka H, Uehara K, Kuwata S, Koiwai A, Noma M. Broad resistance to plant viruses intransgenic plants conferred by anti- sense inhibition of a host gene essential in S-adenosylmethionine-dependent transmethylation reactions. Proceedings. National Academy of Sciences. United States of America 1995;**92**:6117-6121

[142] Bandurska H, Stroinski A, Kubis J. The effect of jasmonic acid on the accumulation of ABA, proline and spermidine and its influence on membrane injury under water deficit in two barley genotypes. Acta Physiologiae Plantarum. 2003;**25**:279-285

[143] Nishiyama R, Watanabe Y, Fujita Y, Le DT, Kojima M, Werner T, et al. Analysis of cytokinin mutants and regulation of cytokinin metabolic genes reveals important regulatory roles of cytokinins in drought, salt and abscisic acid responses, and abscisic acid biosynthesis. Plant Cell. 2011;**23**: 2169-2183

[144] Qiu Z, Guo J, Zhu A, Zhang L, Zhang M. Exogenous jasmonic acid can enhance tolerance of wheat seedlings to salt stress. Ecotoxicology and Environmental Safety. 2014;**104**:202-208

[145] Xie M, Chen H, Huang L, O'Neil RC, Shokhirev MN, Ecker JR. A B-ARR-mediated cytokinin transcriptional network directs hormone cross-regulation and shoot development. Nature Communications. 2018;**9**:1604

[146] Hutchison CE, Li J, Argueso C, Gonzalez M, Lee E, Lewis MW, et al. The Arabidopsis histidine phosphotransfer proteins are redundant positive regulators of cytokinin signaling. Plant Cell. 2006;**18**:3073-3087

[147] Yan Z, Liu X, Ljung K, Li S, Zhao W, Yang F, et al. Type B response regulators act as Central integrators in transcriptional control of the auxin biosynthesis enzyme TAA1. Plant Physiology. 2017;**175**:1438-1454

[148] Argueso CT, Ferreira FJ, Kieber JJ. Environmental perception avenues: The interaction of cytokinin and environmental response pathways. Plant, Cell & Environment. 2009;**32**: 1147-1160

[149] Cortleven A, Leuendorf JE, Frank M, Pezzetta D, Bolt S, Schmülling T. Cytokinin action in response to abiotic and biotic stresses in plants. Plant, Cell & Environment. 2019;**42**:998-1018

[150] Jang G, Chang SH, Um TY, Lee S, Kim J-K, Do Choi Y. Antagonistic interaction between Jasmonic acid and cytokinin in xylem development. Scientific Reports. 2017;7:10212

[151] Tran L-SP, Urao T, Qin F, Maruyama K, Kakimoto T, Shinozaki K, et al. Functional analysis of AHK1/ ATHK1 and cytokinin receptor histidine kinases in response to abscisic acid, drought, and salt stress in Arabidopsis. Proc. Natl. Acad. Sci. USA. 2007;**104**:20623-20628

[152] Choi J, Huh SU, Kojima M, Sakakibara H, Paek K-H, Hwang I. The cytokinin-activated transcription factor ARR2 promotes plant immunity via TGA3/NPR1-dependent salicylic acid signaling in Arabidopsis. Developmental Cell. 2010;**19**:284-295

[153] Igari K, Endo S, Hibara K-I, Aida M, Sakakibara H, Kawasaki T, et al. Constitutive activation of a CC-NB-LRR protein alters morphogenesis through the cytokinin pathway in Arabidopsis. The Plant Journal. 2008;**55**:14-27

[154] Liang YS, Ermawati N, Cha J-Y, et al. Overexpression of an AP2/ERF-type transcription factor CRF5 confers pathogen resistance to arabidopsis plants. Journal of Korean Society for Applied Biological Chemistry. 2010;**53**:142-148. DOI: 10.3839/ jksabc.2010.024

[155] Dervinis C, Frost CJ, Lawrence SD, Novak NG, Davis JM. Cytokinin primes plant responses to wounding and reduces insect performance. Journal of Plant Growth Regulation. 2010;**29**:289-296

[156] Sano H, Seo S, Koizumi N, Niki T, Iwamura H, Ohashi Y. Regulation of cytokinins of endogenous levels of Jasmonic and salicylic acids in mechanically wounded tobacco plants. Plant & Cell Physiology. 1996;**37**: 762-769

[157] Dermastia M, Ravnikar M, Vilhar B, Kovac M. Increased level of cytokinin ribosides in Jasmonic acid-treated potato (*Solanum tuberosum*) stem node cultures. Physiologia Plantarum. 1994;**92**:241-246

[158] Wang W, Barnaby JY, Tada Y, Li H, Tör M, Caldelari D, et al. Timing of plant immune responses by a Central circadian regulator. Nature. 2011;**470**:110-114

[159] Werner T, Nehnevajova E, Köllmer I, Novák O, Strnad M, Krämer U, et al. Root-specific reduction

of cytokinin causes enhanced root growth, drought tolerance, and leaf mineral enrichment in Arabidopsis and tobacco. Plant Cell. 2010;**22**:3905-3920

[160] Reusche M, Klaskova J, Thole K, Truskina J, Novak O, Janz D, et al. Stabilization of cytokinin levels enhances Arabidopsis resistance against *Verticillium longisporum*. Molecular Plant-Microbe Interactions. 2013;**26**(8):850-860

[161] Kohli A, Sreenivasulu N, Lakshmanan P, Kumar PP. The phytohormone crosstalk paradigm takes center stage in understanding how plants respond to abiotic stresses. Plant Cell Reports. 2013;**32**(7):945-957

[162] Vishal B, Kumar PP. Regulation of seed germination and abiotic stresses by gibberellins and abscisic acid. Frontiers in Plant Science. 2018;**9**(838):1-15

[163] Yamaguchi S. Gibberellin metabolism and its regulation. Annual Review of Plant Physiology. 2008;**59**:225-251

[164] Sponsel VMP, . Gibberellin, biosynthesis and inactivation. In: Davies PJ, editor. Plant Hormones Biosynthesis, Signal Transduction, Action! Dordrecht: Springer; 2004. pp. 63-94

[165] Fleet CM, Sun TP. A DELLAcate balance: The role of gibberellin in plant morphogenesis. Current Opinion in Plant Biology. 2005;**8**:77-85

[166] Jiang C, Fu X. GA action: turning on de-DELLA repressing signaling. Current Opinion in Plant Biology. 2007;**10**:461-465

[167] Navarro L, Bari R, Achard P, Lison P, Nemri A, Harberd NP, et al. DELLAs control plant immune responses by modulating the balance of jasmonic acid and salicylic acid signaling. Current Biology. 2008;**18**:650-655

[168] Hou X, Lee LYC, Xia K, Yan Y, Yu H. DELLAs modulate jasmonate signaling via competitive binding to JAZs. Developmental Cell. 2010;**19**:884-894

[169] Yang D-L, Yao J, Mei C-S, Tong X-H, Zeng L-J, Li Q, et al. Plant hormone jasmonate prioritizes defense over growth by interfering with gibberellin signaling Cascade. Proc. Natl. Acad. Sci. USA. 2012;**109**:E1192-E1200

[170] Um TY, Lee HY, Lee S, Chang SH, Chung PJ, Oh K-B, et al. JASMONATE ZIM-DOMAIN PROTEIN 9 interacts with SLENDER RICE 1 to mediate the antagonistic interaction between jasmonic and gibberellic acid signals in rice. Frontiers in Plant Science. 2018;**9**:1866

[171] Huot B, Yao J, Montgomery BL, He SY. Growth–defense tradeoffs in plants: A balancing act to optimize fitness. Molecular Plant. 2014;**7**(8):1267-1287. DOI: 10.1093/mp/ssu049

[172] Wild M, Davière J-M, Cheminant S, Regnault T, Baumberger N, Heintz D, et al. The Arabidopsis DELLA RGA-LIKE3 is a direct target of MYC2 and modulates jasmonate signaling responses. Plant Cell. 2012;**24**:3307-3319

[173] Sasse JM. Physiological actions of brassinosteroids: An update. Journal of Plant Growth Regulation. 2003;**22**: 276-288

[174] Vardhini BV, Anuradha S, Rao SSR. Brassinosteroids: A great potential to improve crop productivity. Indian Journal of Plant Physiology. 2006;**11**:1-12

[175] Bajguz A, Hayat S. Effects of brassinosteroids on the plant responses to environmental stresses, Plant Physiol. The Biochemist. 2009;**47**:1-8

[176] Belkhadir Y, Jaillais Y, Epple P, Balsemão-Pires E, Dangl JL, Chory J.

Brassinos- teroids modulate the efficiency of plant immune responses to microbe- associated molecular patterns. Proc Nat Acad Sci. 2012;**109**(1):297-302

[177] Denancé N, Sánchez-Vallet A, Goffner D, Molina A. Disease resistance or growth: the role of plant hormones in balancing immune responses and fitness costs. Frontiers in Plant Science. 2013;**4**:155

[178] Gómez-Gómez L. Plant perception systems for pathogen recognition and defense. Molecular Immunology. 2004;**41**:1055-1062. DOI: 10.1016/j.molimm.2004.06.008

[179] Zipfel C, Kunze G, Chinchilla D, Caniard A, Jones JDG, Boller T, et al. Perception of the bacterial PAMP EF-Tu by the receptor EFR restricts Agrobacterium-mediated transformation. Cell. 2006;**125**:749-760. DOI: 10.1016/j.cell.2006.03.037

[180] Chisholm ST, Coaker G, Day B, Staskawicz BJ. Host-microbe interactions: shaping the evolution of the plant immune response. Cell. 2006;**124**:803-814

[181] Boch J, Scholze H, Schornack S, Landgraf A, Hahn S, Kay S, et al. Breaking the code of DNA binding specificity of TAL- type III effectors. Science. 2009;**326**:1509-1512

[182] Bogdanove AJ, Schornack S, Lahaye T. TAL effectors: Finding plant genes for disease and defense. Current Opinion in Plant Biology. 2011;**13**: 394-401

[183] Moscou MJ, Bogdanove AJ. A simple cipher governs DNA recognition by TAL effectors. Science. 2009; **326**:1501

[184] Marois E, Vanden Ackerveken G, Bonas U. The Xanthomonas type III effector protein AvrBs3 modulates plant gene expression and induces hyper trophy in the susceptible host. Molecular Plant-Microbe Interactions. 2002;**15**:637-646

[185] Kay S, Hahn S, Marois E, Hause G, Bonas U. A bacterial effector acts as a plant transcription factor and induces a cell size regulator. Science. 2007;**318**:648-651

[186] Chen Z, Kloek AP, Cuzick A, Moeder W, Tang D, Innes RW, et al. The Pseudomonas syringae type III effector AvrRpt2 functions downstream or independently of SA to promote virulence on Arabidopsis thaliana. The Plant Journal 2004;**37**:494-504

[187] Sugio A, MacLean AM, Kingdom HN, Grieve VM, Manimekalai R, Hogenhout SA. Diverse targets of phytoplasma effectors: from plant development to defense against insects. Annual Review of Phytopathology. 2011;**49**:175-195

[188] Hoshi A, Oshima K, Kakizawa S, Ishii Y, Ozeki J, Hashimoto M, et al. A unique virulence factor for proliferation and dwarfism in plants identified from a phytopathogenic bacterium. Proceedings. National Academy of Sciences United States of America. 2009;**106**:6416-6421

[189] Flor HH. Current status of the gene-for-gene concept. Annual Review of Phytopathology. 1971;**9**:275-296

[190] Greenberg JT, Yao N. The role and regulation of programmed cell death in plant–pathogen interactions. Cellular Microbiology. 2004;**6**:201-211

[191] Hammond-Kosack KE, Jones JDG. Resistancegene-dependent plant defense responses. Plant Cell. 1996;**8**:1773-1791

[192] Balbi V, Devoto A. Jasmonate signaling network in Arabidopsis thaliana: crucial regulatory nodes and new physiological scenarios. The New Phytologist. 2008;**177**:301-318

[193] Fonseca S, Chico JM, Solano R. The jasmonate pathway: The ligand, the receptor and the core signaling module. Current Opinion in Plant Biology 2009;**12**:539-547

[194] GGfeller A, Liechti R, Farmer EE. Arabidopsis jasmonate signaling pathway. Sci Signal. 2010;**3**(109):cm4

[195] Doares SH, Narváez-Vásquez J, Conconi A, Ryan CA. Salicylic acid inhibits synthesis of proteinase inhibitors in tomato leaves induced by system in and Jasmonic acid. Plant Physiology. 1995;**108**:1741-1746

[196] Petersen M, Brodersen P, Naested H, Andreasson E, Lindhart U, Johansen B, et al. Arabidopsis MAP kinase4 negatively regulates systemic acquired resistance. Cell. 2000;**103**:1111-1120

[197] Khan N, Bano A, Ali S, Babar MA. Crosstalk amongst phytohormones from planta and PGPR under biotic and abiotic stresses. Plant Growth Regulation. 2020;**90**:189-203

[198] Liu H, Timko MP. Jasmonic acid signaling and molecular crosstalk with other phytohormones. International Journal of Molecular Sciences. 2021;**22**:2914. DOI: 10.3390/ijms22062914

[199] Liu L, Sonbol F-M, Huot B, Gu Y, Withers J, Mwimba M, et al. Salicylic acid receptors activate jasmonic acid signalling through a non-canonical pathway to promote effector-triggered immunity. Nature Communications. 2016;7:13099

[200] Pacifici E, Polverari L, Sabatini S. Plant hormone cross-talk: The pivot of root growth. Journal of Experimental Botany. 2015;**66**:1113-1121

[201] Weiss D, Ori N. Mechanisms of cross talk between gibberellin and other hormones. Plant Physiology. 2007;**144**:1240-1246

[202] Hu Y, Jiang Y, Han X, Wang H, Pan J, Yu D. Jasmonate regulates leaf senescence and tolerance to cold stress: Crosstalk with other phytohormones. Journal of Experimental Botany. 2017b;**68**:1361-1369

[203] Gimenez-Ibanez S, Solano R. Nuclear jasmonate and salicylate signaling and crosstalk in defense against pathogens. Frontiers in Plant Science. 2013;**4**:72. DOI: 10.3389/fpls.2013.00072

[204] Seo PJ, Lee AK, Xiang FN, Park CM. Molecular and functional profiling of Arabidopsis pathogenesis-related genes: Insights into their roles in salt response of seed germination. Plant & Cell Physiology. 2008;**49**:334-344

[205] Caarls L, Pieterse CMJ, Van Wees SCM. How salicylic acid takes transcriptional control over jasmonic acid signaling. Frontiers in Plant Science 2015;**6**(170):1-11

[206] Pieterse CM, Pierik R, VanWees SC. Different shades of JAZ during plant growth and defense. The New Phytologist. 2014;**204**:261-264

[207] Hartweck LM. Gibberellin signaling. Planta. 2008;**229**:1-13

[208] Williams C, Fernandez-Calvo P, Colinas M, Pauwels L, Goossens A. Jasmonate and auxin perception: How plants keep F-boxes in check. Journal of Experimental Botany. 2019;**70**:3401-3414

[209] Heinrich M, Hettenhausen C, Lange T, Wunsche H, Fang J, Baldwin IT, et al. High levels of jasmonic acid antagonize the biosynthesis of gibberellins and inhibit the growth of Nicotiana attenuata stems. The Plant Journal. 2013;**73**(4):591-606

[210] Birkenbihl RP, Diezel C, Somssich IE. Arabidop- sis WRKY33 is a key transcriptional regulator of

hormonal and metabolic responses toward *Botrytis cinerea* infection. Plant Physiology. 2012;**159**:266-285

[211] Li J, Brader G, Palva ET. The WRKY70 transcription factor: a node of convergence for jasmonate-mediated and salicylate-mediated signals in plant defense. The Plant Cell. 2004;**16**(2): 319-331

[212] Zhang X, Wang C, Zhang Y, Sun Y, Mou Z. The Arabidop- sis mediator complex subunit 16 positively regulate ssalicylate-mediated systemic acquired resistance and jasmonate/ethylene-induced defense pathways. Plant Cell. 2012;**24**:4294-4309

[213] Uppalapati SR, Ayoubi P, Weng H, Palmer DA, Mitchell RE, Jones W, et al. The phyto toxin coronatine and methyl jasmonate impact multiple phyto-hormone pathways in tomato. The Plant Journal. 2005;**42**:201-217

[214] Zheng XY, Spivey NW, Zeng W, Liu PP, Fu ZQ, Klessig DF, et al. Coronatine promotes Pseudomonas syringae virulence in plants by activating a signaling Cascade that inhibits salicylic acid accumulation. Cell Host & Microbe. 2012;**11**:587-596

[215] Brooks DM, Bender CL, Kunkel BN. The Pseudomonas syringae phytotoxin coronatine promotes virulence by overcoming salicylic acid-dependent defences in *Arabidopsis thaliana*. Molecular Plant Pathology. 2005;**6**:629-639

[216] Lorenzo O, Piqueras R, Sanchez-Serrano JJ, Solano R. ETHYLENE RESPONSE FACTOR1 integrates signals from ethylene and jasmonate pathways in plant defense. The Plant Cell. 2003;**15**(1):165-178

[217] Song S, Huang H, Gao H, Wang J, Wu D, Liu X, et al. Interaction between MYC2 and ETHYLENE INSENSITIVE3 modulates antagonism between jasmonate and ethylene signaling in Arabidopsis. The Plant Cell. 2014;**26**(1):263-279

[218] Kempel A, Schädler M, Chrobock T, Fischer M, van Kleunen M. Tradeoffs associated with constitutive and induced plant resistance against herbivory. Proceedings National Academy of Sciences United States of America. 2011;**108**:5685-5689

[219] Manzaneda AJ, Kasavajhala VSK, Mitchell-Olds T. Variation and fitness costs for tolerance to different types of herbivore damage in Boechera stricta genotypes with contrasting glucosinolate structures. The New Phytologist. 2010;**188**:464-477

[220] Walters D, Heil M. Costs and trade-off associated with induced resistance. Physiological and Molecular Plant Pathology. 2007;**71**:3-17

[221] Conrath U. Molecular aspects of defence priming. Trends in Plant Science. 2011;**16**:524-531

[222] Latzel V, Zhang Y, Karlsson Moritz K, Fischer M, Bossdorf O. Epigenetic variation in plant responses to defence hormones. Annals of Botany. 2012;**110**:1423-1428

[223] Holeski LM, Jander G, Agrawal AA. Transgenerational defense induction and epigenetic inheritance in plants. Trends in Ecology & Evolution. 2012;**27**:618-626

[224] Robert-Seilaniantz A, Grant M, Jones JD. Hormone crosstalk in plant disease and defense: more than just jasmonate-salicylate antagonism. Annual Review of Phytopathology. 2011;**49**:317-343

[225] Thaler JS, Humphrey PT, Whiteman NK. Evolution of jasmonate and salicylate signal crosstalk. Trends in Plant Science. 2012;**17**:260-270

Chapter 3

# Plant-Microbe Interaction: Prospects and Applications in Sustainable Environmental Management

*Ajuzieogu Chinyere Augusta, Ehis-Eriakha Chioma Bertha and Akemu Stephen Eromosele*

## Abstract

Plant-microbe interaction is mostly mutualistic although sometimes it can be negative. These interactions contribute to improving the environmental quality and health of all organisms. One significant aspect to this is application in sustainable environmental management. Plants are known to be involved in remediation of polluted environments through a mechanism known as phytoremediation and this process is usually more effective in collaboration with microorganism resident within the plant environment. These plants and microbes possess attributes that makes them great candidates for sustainable remediation of impacted environments. Different organic pollutants have been decontaminated from the environment using the phytoremediation approach. The plant-associated microbes possess certain traits that exert selective effect on the growth of plants which consequently perform the decontamination process through different mechanisms. Also, these microorganisms' harbour requisite genes charged with the responsibility of mineralization of different organic and inorganic compounds through several pathways to produce innocuous by-products. The limitations associated with this approach that prevents full-scale application such as contaminant-induced stress frequently leads to low/slow rates of seed germination, plant development and decreases in plant biomass have been solved by using plant growth promoting rhizobacteria. Phytoremediation is an emerging, cost-effective, eco-friendly and operational technology for the cleanup of polluted environment.

**Keywords:** environment, phytoremediation, pollution, plant, rhizobacteria

## 1. Introduction

The idea of phytoremediation of xenobiotics was birthed a few decades back as a result of the awareness that plants possess the capability of metabolizing toxic compounds. Since then, phytoremediation was adjudged a proven technology for the decontamination of environments polluted by a different of organic compounds such as polychlorinated biphenyls (PCBs), pesticides, polyaromatic hydrocarbons (PAHs), chlorinated solvents, dioxins and different approaches like rhizoremediation, combination of PGPR and specific contaminant-degrading bacteria, genetically engineered microbes, transgenic plants and enzyme technology can be used to improve the efficiency of bioremediation. Phytoremediation is an emerging technology that uses plants and associated bacteria for the treatment of contaminated environments by toxic pollutants. Although some challenges that have so far prevented full- scale application of phytoremediation technologies is that contaminant-induced stress frequently leads to low rates of seed germination, slow rates of plant development and decreases in plant biomass. However, this problem can be solved by using plant growth promoting rhizobacteria. Rhizobacteria that exert beneficial effect on the plant growth and development are called as plant growth promoting rhizobacteria (PGPR). The term rhizoremediation has been used to describe the combination of phytoremediation and bioaugmentation with contaminant

Petroleum is composed of various hydrocarbons including aromatics (e.g. polycyclic aromatic hydrocarbons [PAHs]), asphaltenes, aliphatics [i.e. n-alkanes (linear), branched, saturated & unsaturated], & others, which differ in chemical & physical composition based on the reservoir's origin [1, 2]. Petroleum hydrocarbons (PHCs) are organic in nature, comprising of hydrogen and carbon, and highly hydrophobic. In recent times, anthropogenic activities including industrial actions, petroleum & its products (fuel, diesel, kerosene and others) and partial oxidation of fossil fuels, have lead to build-up of PHCs in the environment [3–5]. Actually, petroleum & its products have impacted considerably, on both aquatic and terrestrial ecosystems contaminated by it. Owing to the fact that microorganisms participate directly in bio-geochemical cycles as major players of carbon and petroleum hydrocarbon breakdown (degradation), it is important to further understand petroleum biological degradation and its application [6, 7].

Microorganisms are capable of breaking down or producing hydrocarbons based on some kind of metabolic pathways, unique to each function in the underlying environment [7, 8]. There are key players in the biological degradation of several petroleum products like benzene, toluene, ethylbenzene & xylenes (BTEX), aliphatic & PAHs [2, 9, 10]. Efficient indicators of soil contamination levels are soil microorganisms, plants & biota. They have the ability to breakdown or retain approx. 100% of all kinds of soil pollutants & prevent them from gaining entrance to larger environment [11]. But, when there is severe pollution or contamination of the soil, it results in adverse impacts on soil biodiversity & soil quality. Soil functions (such as, fiber, food and fuel production) is also destroyed, and immediately the food chain is affected, this becomes a threat to public health [9, 12, 13]. Recording success in bio-degradation & bio-transformation of organics in PHC-contaminated soils to less-toxic by-products is usually dependent on the potential to build & sustain conditions that will favour & support microbe-driven degradation, bio-technologically & naturally. The interactions between microbes & plants in phyllosphere & soil are highly significant for both plant growth & productivity in agro-systems, natural systems &/or microbe-driven breakdown of soil PHCs. In field studies, the successful use of plant-microbe interactions in the biological remediation of PHC-contaminated soil relies basically on the native (autochthonous) microorganisms (rhizospheric & endophytic bacteria associated with plant) with the genes unique for bio-degradation of petroleum hydrocarbons [10].

Increasing research on ways to remediate or clean up contaminated environments is the outcome of increased awareness of the danger of soil pollution and its adverse effects on the entire ecological chain. Owing to the diverse nature of pollutants, no absolute remedy exists that is common to all kinds of soil contaminants. In this review however, the pollutants of interest are PHCs sourced from crude-oil or refined petroleum by-products [11]. Thus, researches related to soil HC-contamination & its bio-remediation is the focus. In addition, plant–microbe interactions related to bio-degradation of PHCs could offer a robust understanding of the requisite tools for developing on-site biological remediation plans for mitigating risks in PHCs-contaminated soil.

Terrestrial (soil) pollution can be restored by developing new, science-based technique, including a new emerging method, i.e., bio-remediation. Bio-remediation is an environmentally- friendly & efficient technique, where live microbe(s) and its products or other biological agents (such as plants) can be utilized for the remediation of eco-contamination [14].

The fate of hydrocarbon/organic contaminants in soil-plant environment is determined by significant processes driven by plant-microbe interactions, [15, 16], competence of the microbial activity & microbe-degradation or bio-transformation of petroleum hydrocarbons in soil. In field studies, in depth understanding of the fate of a hydrocarbon/organic-pollutant in oil would aid in determining if the contaminant will persist in the environment or not, enhance the success of any remediation strategy & assist in developing a high-throughput risk mitigation approach.

This part of the review focuses on role of plants & microbes in bio-degradation of PHCs-contaminated soil, resulting from increased researches on bio-remediation & field trials. Also, the enzymatic activities of hydrocarbon degrading microorganisms will be discussed. Emphasis is placed on Phyto-remediation, a valuable method that depends on plants to eliminate/decontaminate soil contaminants. Studies that recorded success on the use of trees in the restoration of PHCs-contaminated soils are also cited.

## 2. Biological remediation

Biological remediation involves elimination, attenuation or transformation of pollutants or contaminants by using biological agents/processes. It is frequently used in moderately PHC-contaminated soils. It is a better remediation tool in soil contamination when compared to physico/chemical remediation. It also offers a cost-effective option, a potentially low-technology, low risk of secondary pollution, & aesthetic value (by phytoremediation) [16–19]. For the past years, the application of bio-degradation and/ or bio-remediation as a remediation/clean-up strategy has become the strategy of choice for remediating PHC-contaminated soil, for the following reasons that; it is cost-effective, sustainable & can enhance natural bio-degradation processes by optimizing limiting factors [9, 20].

Four types of biological strategies useable in soil remediation; (i) micro-organisms (such as bacteria or fungi) to degrade organo-pollutants (also referred to as microbial remediation), (ii) fast growing plants with large biomass, or plant & associated rhizospheric microbial population assisted remediation also known as phyto/rhizo-remediation, (iii) animals in soil (such as worms) to concentrate or stabilize contaminants that cannot be broken down by biological processes, in their body or in the soil; (iv) the combined utilization of the entire strategies aforementioned or the combination of physico/chemical & biological methods. Nevertheless, the focus here is on the use of plants, also known as phyto-remediation [16, 19].

Phyto-remediation is a known bio-remediation technique that uses the degradation potential of microbes & plants to clean up or decrease soil pollutant concentration to permissible risk-levels of site owners and/or regulatory bodies [16, 21–24].

### 2.1 Phytoremediation

Plants have the capacity to adjust to and modify diverse environmental conditions to certain levels [25]. Phyto-remediation (phyto—Greek for plants) is a universal word that refers to diverse methods that employ plants in the clean-up of environmental (water & soil) contaminants [16, 26, 27]. It is the use of living green plants for on-site remediation. It also involves taking advantage of the symbiotic interaction between plant-based processes & their associated microbial communities to eliminate, transform, &/or mineralize soil inorganic & organic contaminants, as well as pollutants in wastewater, surface water, ground water & sludges [28–30]. In specific words, phytoremediation is a term used to describe a battery of technologies that utilize plants to reduce, remove, degrade or immobilize environmental toxicants/pollutants with the sole purpose of eco-restoration (restoring a site to a condition useable for private or public applications) [16, 31] as described in (**Figure 1** and **Table 1**). Phytoremediation has been widely employed to eco-restoration of soil contaminants such as landfill leachates, crude-oil, metals, pesticides, solvents, explosives, etc. [42, 43]. Phylogenetic diversity of PHC-degraders is huge & numerous recurrent groups reported by most phyto-remediation studies are; *Stenotrophomonas, Sphingomonas, Rhodococcus, Acinetobacter, Alcaligenes, Arthrobacter, Burkholderia, Flavobacterium, Mycobacterium, Micrococcus, Nocardioides, Pseudomonas,* & *Ralstonia* species [10, 44–46].

Individual mechanisms described above are known to have major impact on concentration, environmental outcome & behavior, toxicity, bio-availability & bio-accessibility of PHCs in contaminated soil. Field studies have shown that, certain plants have the ability to eliminate/breakdown xenobiotic organic compounds from the environment by enhancing accumulation & transformation [16, 47], extracellular transformation [48, 49], & metabolic activities of microbes around rhizosphere of the plant [16, 40]. Strategies for phyto-remediation of foreign organics are grouped into two groups; direct phyto-remediation (*in planta*) and phyto-remediation (*ex planta*) [16, 24, 50, 51]. The phyto-remediation *ex planta* depends on a synergy association between substances secreted by roots (root exudates) & metabolic activities of autochthonous rhizosphere related micro-organisms [32, 52]. Plants with the capacity for PHCs phytoremediation have been reported in a number of studies as shown in **Table 2**.

Chitara et al. [24, 54], stated that an efficient plant for phyto-remediation should have these traits; (i) capacity for tolerance, build-up, or breakdown of pollutants in their above-ground areas, (ii) tolerate the volume of pollutants built-up, (iii) grow fast & produce high cell mass, (iv) fibrous root systems, & (v) be easy to harvest. The fibrous root systems are advantageous over taproot systems because they offer a larger surface area for colonization of microbial populations & also enhance the interaction between autochthonous rhizospheric-associated microbial communities & the xenobiotic compounds [16, 55].

Aside *in vitro* trials, success reports have been recorded with phyto-remediation field trials aimed at remediating PHC–contaminated soil over 2 decades ago.

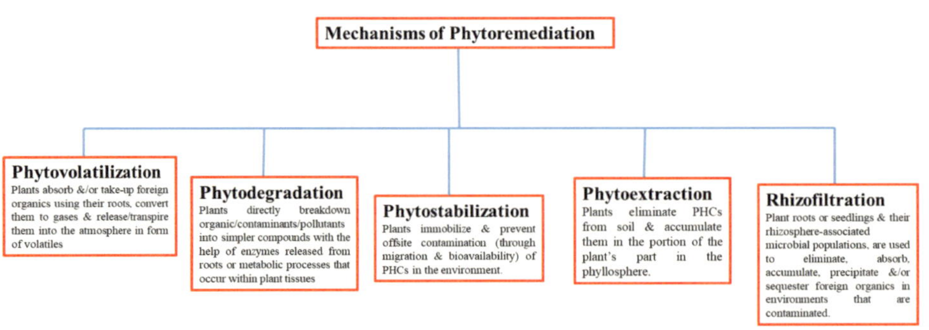


**Figure 1.**
*Mechanisms involved in phytoremediation.*

DOI: http://dx.doi.org/10.5772/intechopen.102690

| Position | Mechanisms | Description | Pollutants and media | Objectives of phyto-remediation | References |
|---|---|---|---|---|---|
| Plant shoots | Phyto-accumulation; Phyto-sequestration; Phyto-extraction | Plants eliminate PHCs from soil & accumulate them in the portion of the plant's part in the phyllosphere | Metals, PHCs & other inorganic toxicants present in sediment, surface water & soil | Phyto-remediation is achieved by removing plants that have accumulated the pollutant | [32, 33] |
| | Phyto-transformation or Phytodegradation | Plants directly breakdown organic contaminants/pollutants into simpler compounds with the help of enzymes released from roots or metabolic processes that occur within plant tissues, which consequently sustains growth of plant | PHCs, mobile organics such as herbicides in surface water, sediment & soil | Phyto-remediation by complete mineralization | [32, 34] |
| | Phyto-hydraulics | Plants are used to increase evapo-transpiration, thus, putting the movement of contaminant, water & soil under control | Inorganics & Organics in surface & groundwater | Plants contain the pollutant by controlling water movement | [35, 36] |
| | Phyto-volatilization | Plants absorb &/or take-up foreign organics using their roots, convert them to gases & release/transpire them into the atmosphere in form of volatiles | Volatile organics & in-organics such as mercury & selenium in soil surface & surface water | Biological remediation by plant removal | [37, 38] |
| Rhizosphere | Rhizo-filtration | Plants (aquatic/terrestrial), their roots or seedlings (rhizo-filtration or blasto-filtration), & their rhizosphere-associated microbial populations, are used to eliminate, absorb, accumulate, precipitate &/or sequester foreign organics in environments that are contaminated | Inorganics such as heavy metals & Organics in surface water | Containment | [39] |
| | Phyto-immobilization or Phyto-stabilization | Plants immobilize & prevent offsite contamination (through migration & bioavailability) of PHCs in the environment | Inorganics & organics in water & soil | Containment | [39] |
| | Phyto-stimulaton or Rhizo-degradation | Plant-assisted bio-remediation which basically depends on degradation of contaminants via metabolic activity of microbes (fungi, yeast, or bacteria) in soil | Hydro-phobic organics such as Polychlorinated biphenyls, PAHs, & other PHCs in water & soil | Remediation by mineralization | [40, 41] |

**Table 1.**
*Mechanisms involved in phyto-remediation.*

| S/N | Botanical name | Common name |
|---|---|---|
| 1. | *Agropyron smithii* | Western wheat grass |
| 2. | *Andropogon geradi* | Big bluestem |
| 3. | *Bassia scoparia L.* | Burning bush or ragweed |
| 4. | *Biden* | Beggar ticks |
| 5. | *Bouteloua gracilis* | Blue grama |
| 6. | *Cynodon dactylon* | Bermuda grass |
| 7. | *Glycine max* | Soybean |
| 8. | *Lolium perenne L.* | Ryegrass |
| 9. | *Medicago sativa L.* | Alfalfa |
| 10. | *Oryza sativa or Oryza glaberrima* | Rice |
| 11. | *Zea mays L.* | Maize |
| 12. | *Sorghum bicolor* | Sorghum |
| 13. | *Boutelova curtipendula* | Side oats grama |
| 14. | *Sorghastrum nutans* | Indian grass |
| 15. | *Fetusca rubra var. arctared* | Arctared red fescue |
| 16. | *Daucas carota* | Carrot |
| 17. | *Sorghum vulgare L.* | Sudan grass |

*Adapted from [29, 53].*

**Table 2.**
*Some plants with phyto-remediation ability for the clean-up of petroleum hydrocarbon–contaminated soil.*

Cook and Hesterberg [56] stated that trees & grasses are usually employed for phyto-remediation purposes, with trees classically selected for the bio-remediation of Benzene, Toluene, Ethybenzene, Xylene, whereas grasses are often employed for the bio-remediation of PHC–contaminated soil.

Study on plants with the potential for enhancing bio-remediation of PHC-contaminated soil [16, 57], obtained results that indicated that the growth of *Glycine max* impacted on soil organic matter contents, moisture & pH of PHC–contaminated soils significantly, with levels of significance; ($P < 0.001$, $P < 0.01$ & $P < 0.05$).

Biodegradation of PHCs was improved in soils spiked with 25 g/L crude-oil & cultivated with *Glycine max* & the soils became more suited for growth of plants as weeds were observed to also grow from the soil. Findings from the study revealed the farming of certain crops like *G. max*, could be proficient in ecorestoration & alleviate risks of PHC–contaminated soil [57]. Phyto-remediation is an inexpensive plant–based remediation strategy for decontaminating PHCs-contaminated soils particularly in the tropics with low finances [58, 59].

Ecological rehabilitation (restoration of contaminated or degraded soils), with the cultivation of *Vetiveria zizanioide* has been reported to increase biomass greatly & consequently improve phyto-remediation of an oil shale mined land contaminated with heavy metals [42, 60]. *Vetiveria zizanioide* (a hydrophyte & xerophytes), tolerates varying abiotic stresses significantly & has been employed in times past in the rehabilitation of coal, gold mines & mining overburdens [61, 62]. In the study by [62], it was discovered that Goose grass (*Eleusine indica*) improved phyto-remediation greatly in soils polluted with PAHs & Total Petroleum Hydrocarbons (TPH).

Comparatively, grasses exhibit traits of fast growth, strong resistance & large biomass in contaminated environment, efficient stabilization & rehabilitation of contaminated lands in sub-tropics & tropics, than trees & shrubs [63, 64]. The capacity of regular tropical grasses, such as elephant grass (*Pennisetum purpureum*), to increase the degradation of a crude-oil contaminated soil has been documented [64]. Field trials have shown that phyto-remediation can be used on soils with moderate PHC contamination or after the use of other biological remediation strategies, in order to alleviate risks linked with PHC-residues in soil further [16, 65]. In a way to improve this bio-remediation strategy, the competence of the plant for phyto-remediation could be significantly enhanced using genetic engineering technologies [24, 66].

Phytoremediation is promising for the onsite treatment of PHC-polluted soils. Treating on site could be challenging to regulate than off-site treatments, for instance, *ex situ* treatment of soil impacted with wastes from refinery. In spite of this, remediation actions on site are largely employed in recent times since they are inexpensive and prevent disruption of contaminated soils. Success of biological degradation is affected by several environmental parameters; contaminants composition, concentration & bio-availability, soil nutrients, oxygen, moisture content, pH, & profile of contaminated area [2, 67, 68]. Understanding ways of controlling these parameters/factors in order to optimize biological activities that will result in bioremediation is paramount.

### 2.2 Choice of species in phyto-remediation

#### *2.2.1 Plant choice*

Several criteria should be considered before choosing plants for phytoremediation. A plant species for phytoremediation should have roots that can spread throughout the whole contaminated site. The principle for plant selection ought to follow the needs of the use, contaminant type & their ability to grow & increase on contaminated soil [16, 18]. Indigenous plants are preferable, to prevent introducing invasive species. E.g. *Hibiscus Cannabinus* (Kenaf) & *Vetivera zizanioides* (Vertiver) which are indigenous plants have been reported to be very proficient in crude-oil pollution remediation in Nigeria [42, 69].

Herbaceous plants, deciduous trees & conifers are renowned plant types [65], based on the environmental conditions & the polluting compound. Peas, clover, reed canary grass & alfalfa (legumes), ryegrass, sunflowers & wheatgrass (grasses) and *Thespesia populnea*, *Populus* sp., *Salix* sp., *Scaevola serica*, *Prosopis pallida* & *Cordia subcordata* (trees) have been reported to display tolerance to PHC-contamination [16, 18, 42, 65]. Tolerance refers to the potential of a plant to grow in hydrocarbon-contaminated soil, however, it does not really suggest the healthiness of the plant [65].

i. **Herbaceous plants in phyto-remediation:** Grasses usually cultivated with trees are largely employed as the main remediation species in hydrocarbon-polluted soils, because they make available great fine roots in topsoil. Grasses are successful as binders & transformers of PHCs like PAHs & BTEX because of the extensive fine root biomass that contains vast microbial community than other species of similar capacity [16, 18, 70].

ii. **Legume—rhizobium symbiosis in phytoremediation:** Nutrients (nitrogen & phosphorus) are especially limiting in contaminated soils. Also, competition for nutrients amongst soil organisms makes nutrients a limiting factor for bio-remediation [71, 72]. When soil moisture content & temperature is low, nitrogen insufficiency is worsened owing to poor mobility of nutrients, limited microbial & enzymes activity [71]. Adequate fertilization & frequent tillage were suggested by [73, 74] as helpful measures in ensuring breakdown of PHC in comparison to un-treated soil. In their research, initial concentrations of PHC were eliminated by 70–81% through bio-remediation in fertilized soils as against 56% elimination without fertilization in natural attenuated soil.

Though, disproportionate application of nitrogen-containing fertilizers could lead to environmental pollution/problems. In order to prevent this, nitrogen-fixing plants like legumes, can be used in place of them [16, 75]. Rhizobia have the capacity to infiltrate roots of legumes, forming symbiotic relationships & nodules, which have the ability to fix gaseous nitrogen into plant in the form of ammonia [24, 76]. *Anabaena,* Blue-green algae, *Azotobacter*, *Azospirillum*, *Rhizobium*, *Actinomycetes* & *Frankia* are generally used Nitrogen-fixers in soils [77, 78].

iii. **Trees as phytoremediation tool:** Trees & their hybrids are extensively employed in clean-up of PHC-impacted soils. CLUIN phyto-remediation databank records that a great percentage of phyto-remediation successful studies were performed using trees (**Table 3**). Plant hybrids that grow fast with desired characteristics like resistance to harsh soil & climatic conditions, resistance to pests & diseases; have been chosen as potential phyto-remediation choice species [16, 80]. For example, hybrid trees from willows & poplars have generally & successfully employed in phyto-remediation of soils polluted with organic & in-organic compounds. But, precautions are advised, to evade risks of utilizing genetically engineered or modified breeds [29, 42, 80].

### 2.3 Rhizoremediation

Although some studies have endorsed the use of plants only, for effective biological remediation of PHC-impacted soils, [81, 82], using plants associated with PHC–eating microbes &/or plant growth–promoting bacteria (PGPB) for the clean-up or degradation of HC-impacted soil has an edge because it reduces the risks of reverse transformation &/or HC-residues [16, 83–85]. The root system of plants which is generally known to offer support & enhance water & nutrient uptake, is a chemical factory that drives several interactions like mutualistic relationships with beneficent autochthonous micro-organisms (e.g. mycorrhizae, endophytes, plant growth-promoting rhizobacteria (PGPR) & rhizobia) below soil surface [16, 24, 86]. The use of phyto-remediation/plant-associated microbes' combination strategy offer better biological clean-up platform than using plant only.

Rhizo-remediation is thus, the use of plants & their interactive relationships with micro-organisms that inhabit the area around the roots (rhizosphere). This combination has the capacity to breakdown foreign organics in the rhizosphere. In this process, root exudates/secretions enhance the survival & metabolic activities of PHC-degrading microbial populations &/or associated rhizo-microbes, resulting in complete breakdown of organics in hydrocarbon-inundated soils [16].

Rhizo-remediation is one of the most efficient phyto-remediation tools that take advantage of the most active area being near/around the roots of plants for removal/degradation/clean-up of organic contaminants [24, 50, 87]. Practically, microbial communities associated with the rhizosphere are the major contributors to biodegradation & the green plants employed are seen as biological, solar-driven pump & treatment systems [43, 88, 89].

The success of rhizo-remediation relies upon the proficiency of the rhizo-microbes, indicated by their potential to survive & compete for root exudates in the rhizosphere, in order for them to be maintained in requisite numbers & proficiently colonize the emerging root system [16, 90]. In field studies, effectiveness

| Name of project (duration) | Trees & other plants employed | Pollutants (initial concentration in media) | Phyto-remediation mechanisms | Results & findings |
|---|---|---|---|---|
| Phyto-remediation at a gasoline release site in Georgia (1999–2002) | Native sedge, Cattails rush, White willow, Black willow, Wooly bull rush | Gasoline in soil and ground water (Soil average BTEX: 1400 μg/L; average benzene: 44 μg/L) | Phyto-degradation; Phyto-volatilization; Rhizo-degradation | 82% reduction in Soil BTEX concentrations; In the 1st year of the growing season, almost 90% of the trees planted survived, although highest death rates was observed in regions with highest concentration of gasoline. In plant branches & leaves, BTEX & benzoic acid (a product of degradation) were present in low concentrations. |
| Phyto-remediation at the Edward Sears Property in New Jersey (1995–2004) | Hybrid poplar | Mixture of organics (e.g. 2700 mg/L of Xylenes) in groundwater & soil | Phyto-degradation; Hydraulic Control | During the 1st 3 years of monitoring, approx. all of the contaminants decreased |
| Phyto-remediation at Oneida Tie Yard Site in Tennessee (1997-) | Hybrid poplar | 17,500 ppb of PAHs & 18,500 ppb of Naphthalene in soil | Rhizo-degradation; Phyto-volatilization | Concentrations of PAHs & naphthalene were 6400 ppb & 4900 ppb respectively, at the end of 7 years monitoring |
| Phyto-remediation at Ashland Inc. in Wisconsin (2000-) | Understory grasses; Hybrid poplar | Diesel in soil, BTEX, gasoline, & other organics in ground water & soil | Rhizo-degradation; Hydraulic Control; Phyto-extraction | Trees tripled in height, & subsurface aeration increased in soil since planting. Roots observed at 10 feet depth during 1st growing season |

*Source: [79].*

**Table 3.**
*Studies that used forest trees in phyto-remediation of PHCs contaminated soils.*

& success in the use of rhizoremediation strategy depends mainly on the potential of PHC-degrading microbial populations &/or PGPR to efficiently colonize the rhizosphere [91].

Employing rhizo-remediation in breakdown of PHCs have been proven to be an inexpensive strategy & it can be further improved by employing genetically modified/engineered microbes &/or plants fashioned uniquely for the purpose & optimizing favorable conditions for efficient restoration/clean-up of organic pollutants [24].

The study by [87] evaluated the plant/associated rhizo-microbe degradation of diesel-polluted soil using two varieties of rapeseed & HC–degrading microbes. They found out that the rapeseeds defenses responded in different ways. Research by [92] revealed that the mean of the residual PAHs in mixtures (48%), was considerably less than PAHs in soils that used plants alone (55%) & in soils that did not employ plants (70%).

## 3. Application of plant growth promoting microbes (PGPM) in phytoremediation

Plant growth-promoting rhizobacteria (PGPR) are a class of beneficial microbes associated with root system (rhizosphere) in plants & on colonization, are known to facilitate plant growth via direct & indirect processes [93, 94] (**Figure 2**). Direct mechanisms include heavy mineral uptake by plants [24, 95], phyto-stimulation (also known as phytohormone production), siderophore production which limits the iron (Fe) activity [24, 96], nitrogen fixation, phosphate solubilization & potassium solubilization; while indirect mechanisms include ISR (induced systemic resistance against plant diseases, also known as "systemic resistance"), phyto-remediation, signal interference [97], antibiotics production, quorum sensing [98], chitinase & glucanase activity, exopolysaccharide production [99]. The PGPM facilitates growth of plants under stress by production of vital enzymes like rhizobi-toxine exopolysaccharides, 1-aminocyclopropane-1-carboxylate (ACC)-deaminase & chitinase.

Plant growth–promoting rhizobacteria (PGPR) can enhance growth of plant in polluted soils by diverse processes or can assist in biological remediation of polluted soil [100] by using any or a combination of the mechanisms stated above. They are capable of detoxifying the contaminated soil by sequestering metal ions inside the cell [101], biotransformation/transformation of metals from toxic to less toxic ones [19, 100, 102], adsorption/desorption of metals, etc. Some examples of PGPR, pollutants they target & the processes they employ to enhance growth of plants under stressed conditions (pollution) are displayed in **Table 4**.

It is worthy to note that this approach applies PGPR whose action is affected by climate change. Thus, successful remediation with PGPR is greatly connected with climate, for e.g. heat could impair plant physiology & growth, likely resulting to modifications in the structure, population, or activity of plant-associated microbes. Therefore, microbial populations with beneficial impacts on plant health or growth may be reduced under unfavourable conditions [24, 111]. Therefore, understand-ing growth patterns of plants & its ambient environment prior to using PGPR is essential, particularly in certain conditions. Identifying a particular PGPR unique to a certain area is thus, essential for achieving improved activity by them & effec-tive enhancement of the bio-remediation of polluted soil under evolving climatic conditions [24, 111].

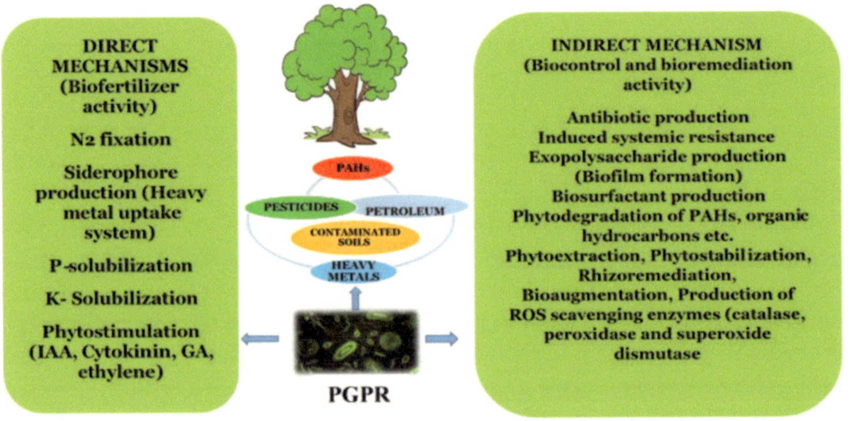


**Figure 2.**
*Mechanisms of action of PGPR in biological remediation of polluted soil.*

| PGPR | Plants | Target pollutant | Mechanism of action | Reference |
|---|---|---|---|---|
| *Dokdonella*, *Luteimonas*, *Sphingomonas*, *Pseudomonas*, & *Sphingobium* sp. | — | Polycyclic aromatic hydrocarbons (PAHs) | Degradation of phenanthrene, fluorene, & pyrene | [103] |
| *Acinetobacter* sp. PDB4 | Rice | Anthracene, Pyrene & Benzo(a)pyrene (BaP) | Siderophore production, phosphate solubilization | [104] |
| *Burkholderia* sp. XTB-5 | *Brassica chinensis*, *Ipomoea aquatic* | Phenol | Siderophore production, Phosphate solubilization & 1-aminocyclopropane-1-carboxylate (ACC) deaminase synthesis | [105] |
| *Pseudomonas plecoglossicida* (JX149549), *P. aeruginosa* (JX100389) | Wheat | Petrol engine oil | Biosurfactant synthesis, petroleum hydrocarbon metabolism, iron sequestration, petroleum hydrocarbons metabolism | [106] |
| *Shinella* sp. EIKU6, *Micrococcus* sp. EIKU8, & *Microbacterium* sp. EIKU5 | — | Arsenic (As) & Uranium (U) | Oxidation & Resistance, Uranium elimination | [107] |
| *Planctomyces Lysobacter*, *Klebsiella* sp. D5A, *Pseudomonas* sp. SB, *Pseudoxanthomonas* | *Testucaarundinacea* | Petroleum hydrocarbons | Production of phyto-hormones & solubilization of minerals; production of biosurfactant; increase root biomass | [108] |
| *Staphylococcus carnosus*, *Bacillus circulans*, & *Enterobacter intermedius*, *Serratia marcescens* BC-3, *Pseudomonas aeruginosa* SLC-2 | Maize and Oat | Petroleum hydrocarbons | Siderophore production, Synthesis of Indole acetic acid, 1-Aminocyclopropane-1-carboxylate (ACC) deaminase activity and petroleum degradation | [93, 109] |
| *Pseudomonas fluorescens ATCC 17400* | Red clover | Radionuclide cesium | Increase in translocator factor, Resorption of Cesium onto biofilms | [110] |

**Table 4.**
*Some PGPR, target pollutant & mechanisms of action.*

## 4. Enzymatic activities of hydrocarbon degrading microbes

High enzymatic potentials present in microbes afford microbial communities the ability to degrade complex hydrocarbons [49, 112]. This petroleum degrading/modifying potential makes them able to breakdown/transform some pollutants like petroleum, & this sums up the significance of enzymes in bio-remediation. The diverse nature of microbial genes adds to the versatility of their metabolic reactions for the transformation of toxicants into less-toxic end-products, subsequently incorporated into natural bio-geochemical cycles [2, 49]. Myriad of micro-organisms (green algae, bacteria, fungi, & cyanobacteria), have PHC-degradative potentials under anaerobic, aerobic, pH, saline & other types of environmental conditions [7, 113]. These enzymatic tools provide these potentials to microbes.

## 4.1 Aerobic degradation of petroleum and petroleum degrading enzymes

Degradation of petroleum is a gradual process that involves sequential breakdown (metabolism) of its components. The genes that encode the production of petroleum degradation enzyme may be found on plasmid or chromosomal DNA [113, 114]. Biological degradation of hydrocarbons can take place under oxic or anoxic conditions [1, 2].

Under oxic conditions, oxygenases introduce oxygen atoms into hydrocarbons (mono-oxygenases introduce one oxygen atom to a substrate while dioxygenases introduce two). Aerobic breakdown of HCs can be quicker, because of $O_2$ (oxygen) available as an electron acceptor [115]. Oxidation of saturates (aliphatics) is acetyl-CoA, usually broken down in the Kreb's cycle, with the synthesis of electrons in the electron transport chain (ETC). The ETC is repeated, breaking down HCs further to carbon dioxide ($CO_2$) [116]. Aromatics like naphthalene & BTEX can also be broken down under oxic conditions. Breakdown of these compounds leads to the first step in catechol synthesis or a similar compound. Once synthesized, catechol could be broken down into precursors in the Kreb's cycle, that are eventually completely mineralized to carbon dioxide ($CO_2$) [115, 116].

### *4.1.1 Alkane degradation*

In recent times, variation in alkane degradation genes clustering & regulation amongst species has been discovered. The finding is that a species could have multiple genes that encodes for various enzymes performing related functions. alkBFGHJKL operon has been reported as one of the operons that encode the enzymes required for alkanes conversion to acetyl-CoA [2, 117].

The reported alk gene products include; AlkS (positive regulator of the alkBFGHIJKL operon & alkST genes), AlkT (rubredoxin reductase), AlkL (outer membrane protein that maybe involved in uptake), AlkK (acyl-CoA synthetase), AlkJ (alcohol dehydrogenase), AlkH (aldehyde dehydrogenase), AlkF & AlkG (rubreoxins), &. AlkB (alkane hydroxylase). These genes have been identified in several petroleum-metabolizing organisms like *Alcanivorax* sp., *Rhodococcus* sp., *Pseudomonas putida*, *Acinetobacter* sp. & others. Andreolli et al. [46, 116, 118] reported thirty-six percent (36%) of the hydrocarbon-metabolizing species obtained in their study possessed genes involved in the metabolism of both n-alkanes (alkB) and aromatic hydrocarbons (xylE). Brzeszcz et al. [2, 113, 116] indicated the coexistence of multiple-degradative potentials in one microorganism (*Pseudomonas* sp. strain BI7), showing both genetic proof and phenotypic responses. Other microbes with similar potentials include members of the genera *Mycobacterium*, *Rhodococcus* and *Pseudomonas*.

Alkane hydroxylases are a class of enzymes that catalyses the breakdown of alkanes & this class of enzymes are present in many diverse bacterial, fungal & algal species [2, 112, 113, 119]. In addition, [117] projected three classes of alkane-degrading enzyme systems; C17+, C5-C16, & C1-C4, which are for degradation of long chain alkanes (broken down basically by unknown enzymes), pentane to hexadecane (broken down by integral membrane cytochrome P450 enzymes or non-heme iron) & methane to butane (broken down by methane-monooxygenase-like enzymes) respectively. The authors also documented bacterial P450 oxygenase system and di-oxygenase (CYP153, class I), eukaryotic P450 (CYP52, class II), alkane hydroxylases related to AlkB genes, the compositions, cofactors, ranges of substrates, presence of the main groups of alkane hydroxylases (soluble methane mono-oxygenase (sMMO), and particulate methane mono-oxygenase (pMMO). They also added that alkane degrading microbes could have multiple alkane hydroxylases, thus have the capacity to metabolize a variety of substrate ranges.

Over the years, amongst the mostly studied alkane degradation pathways is that explained for *Pseudomonas putida* Gpo1, encoded by the OCT plasmid [1, 113, 120, 121]. Converting alkane to an alcohol by this microbe is initiated by a membrane mono-oxygenase, rubredoxin reductase & soluble rubredoxin [1]. van Hamme et al. [1] developed a model for alkane catabolism in Gram-negative bacteria, describing the position & roles of the ALK-gene products. A class of iron-containing enzymes in bacteria called catechol di-oxygenase is an example of those involved in aerobic catabolism of aromatics. They have the ability to hasten the addition of oxygen ($O_2$) atoms to 1,2-dihydroxybenzene (catechol) & its derivatives, with subsequent cleavage of the aromatic ring [2, 113, 115, 116]. Catechol di-oxygenases & similar enzymes involved in cleavage of aromatic ring are accountable for the myriad of microbes with aromatic-HC degrading potential [2, 46, 113, 114, 118].

#### *4.1.2 Polycyclic aromatic hydrocarbon (PAHs) degradation*

In recent times, majority of reports on PAH-degrading genes arise from studies on naphthalene-degrading plasmids like NAH7 from *Pseudomonas putida* strain G7. Naphthalene dioxygenase is now a known versatile enzyme, with the capacity to mediate the catalysis of a broad array of reactions. Genomic & bio-chemical data have proved that enzyme system for naphthalene degradation is also capable of mineralizing other aromatics like anthracene & phenanthrene. A number of other bacteria with PAH-degrading potential have been obtained & characterized. In addition, more genomic tools to study microbial populations have been invented, there by affording researchers the opportunity to realize diversities of PAH metabolic genes [2, 113, 122].

Novel gene sequence & orders have been reported in many species including, *Pseudomonas* sp. strain U2; nagAaGHAbAcAB, phnFECAcAB, *Norcardiodes* sp. strain KP7; phABC *Burkholderia* sp. strain RP007; etc. The ability of several species to degrade a wide range of aromatics is attributed to presence of multiple oxygenases, presence of multiple metabolic pathways or genes, & relaxed initial enzyme specificity for PAHs. The presence of alkane & aromatic hydrocarbon-degrading genes within single species is common [2, 112, 113].

### **4.2 Anaerobic degradation**

Anaerobic degradation is as important as the aerobic degradation process in bioremediation, even though HC-degradation under aerobic process is faster. This could be attributed to several limiting environmental conditions like insufficient oxygen (typical in aquifers, sludge digesters, mangroves) [3, 7]. Anaerobes like sulphate-reducing bacteria catalyse anaerobic degradation using diverse terminal electron acceptors (TEAs) [1, 2, 123]. Usually, anaerobic degradation involves the conversion of aromatics to benzoyl-CoA (target of the benzoyl-CoA reductase (BCR)) action [2, 113, 124]. Environmental conditions determine the TEAs that could be used in the degradation. Fe (III), sulphate & nitrate are examples of TEAs that could be used [2, 72, 113, 115].

Studies have reported that HCs such as toluene, alkylbenzenes (m, o, & p-xylene & trimethylbenzenes), benzene, naphthalene & phenanthrene, > C6 n-alkanes, branched alkanes & HC mixtures can be catabolized under anaerobic conditions. These reactions may take place under denitrifying, sulphate-reducing & Fe (III)-reducing conditions, by anoxygenic photosynthetic bacteria. Sulphate-reducing

*Desulfococcus oleovorans* Hxd3 is the only currently known anaerobic bacterium that degrades n-alkanes independently of anaerobic generation of oxygen species [7, 125].

In recent times, electron acceptors shown to be used during anoxic (anaerobic) degradation of HCs include soil humic acids, manganese oxides, etc. Also, the number of pure cultures shown to catabolize different HCs with various electron acceptors has risen. Examples are members of Proteobacteria group, which have helped in explaining the basic genomic & biochemical processes mediating anoxic breakdown of HCs [2, 46, 125].

The diversity & unique characteristics of anaerobic HC-utilizing bacteria are areas that require more studies. More focus is required on isolating & characterizing enzymes mediating anoxic degradation of HCs especially from non-cultivable organisms (using metagenomics) [2, 7, 113].

Bio-catalysis is creating novel paths geared towards improvement & development of processes & products that will cut down on industrial costs, generation of secondary pollutants (toxic sub-products) & subsequently, the adverse effects on the environment. Enzymatic bio-remediation and creation of new clean energy contributes to reduce harm caused by fossil fuel [13, 113, 126]. Enzyme-mediated remediation can be easier compared to using intact microbes. Some advantages, including the enzymatic potential, can be increased in laboratory conditions [113, 127]. Using enzymes alone does not result to production of toxic by-products [113, 128] and competition from intact cell is not needed [113, 126]. Peixoto et al. [113, 127] stated the major areas to be taken into consideration during enzyme-mediated bio-remediation, range from selection of organisms that has contaminant-degrading potentials, identifying the gene encoding the specific enzyme, to enzyme production.

An example of enzymatic bio-remediation is de-toxification of aromatics (PAHs) & this is can be successful by the application of laccases. Laccases are enzymes that speed up the breakdown (oxidation) of anilines, polyphenols & phenols, with the production of water as the end-product [113, 129, 130]. The major benefit of enzyme-mediated bio-remediation of partially soluble pollutants or hydrophobics (PAHs) is the fact that it can take place in the presence of solvents organic in nature [131]. However, the drawback is that the relevant enzymes could be denatured, inhibited or un-stable in organic solvents. Brzeszcz et al. [2, 131] in their study expressed laccase from *Myceliophthora thermophila* (MtL) in *Saccharomyces cerevisiae*, using directed evolution & extensively improved laccase expression. Years ago, [7, 113, 132] reported success in first field trial with an enzyme-based product, based on the enzyme TrzN, confirming that the enzyme-mediated bio-remediation can effectively clean-up herbicides-contaminated aquatic systems, but only few field studies with enzyme-mediated bio-remediation are available currently. Ismail et al. [7, 133] stated that more than 1000 aromatics-degrading enzymes have been documented.

In 2011 the U.S. EPA outlined 20 bio-remediation agents & one pure enzyme additive alone known as "Petroleum Spill Eater II". The manufacturer described the product as a "bioremediation agent (biological enzyme additive (previously listed as a nutrient additive))," with a 5-year duration [113, 134]. The manufacturer reported 33.6% decrease in aromatics & 36.9% decrease in alkanes after 7days & 89.6 % & 89.8% decrease of the same compounds respectively, 28 days post Petroleum Spill Eater II application, indicating maximum reductions over a short duration.

Although enzyme-mediated bio-remediation is beneficial, there are requirements & challenges which restrict its application to few classes [132]. Generally, these challenges are associated with enzyme stability & high costs.

Genomic techniques are thus extensively being explored, in order to offer enzyme products that can compete favorably as bio-remediation products. Genomic techniques make detection of genes that encode HC-degrading enzymes in environmental samples or micro-organisms possible, thereby acting as high-throughput technique for bio-prospecting studies. In addition, gene-engineering can significantly enhance cost-effective enzyme production [7, 113, 126]. Brzeszcz and Kaszycki [2, 126] reported that new studies using omics (proteomics, protein engineering & metagenomics) are successfully adding to cost-effectiveness, increased cost-benefit ratios & reduction in chemical application. Using genomic techniques for bio-catalysis (bio-degradation) uses can also assist in tackling the challenge of employing genetically modified organisms (GMO) in the environment [113, 126]; for example, recruiting modified microbes into the environment will not be necessary, if modified enzyme is produced in the laboratory.

## Author details

Ajuzieogu Chinyere Augusta[1], Ehis-Eriakha Chioma Bertha[2]*
and Akemu Stephen Eromosele[2]

1 Department of Microbiology, Renaissance University, Enugu State, Nigeria

2 Department of Microbiology, Edo State University Uzairue, Edo State, Nigeria

*Address all correspondence to: bertha_chioma@yahoo.com

## References


[1] van Hamme JD, Singh A, Ward OP. Recent advances in petroleum microbiology. Microbiology and Molecular Biology Reviews. 2003;**67**(4):503-549

[2] Brzeszcz J, Kaszycki P. Aerobic bacteria degrading both n-alkanes & aromatic hydrocarbons: An undervalued strategy for metabolic diversity & flexibility. Biodegradation. 2018;**29**: 359-407

[3] Santos HF, Carmo FL, Paes JE, Rosado AS, Peixoto RS. Bioremediation of mangroves impacted by petroleum. Water, Air, and Soil Pollution. 2011;**216**(1-4):329-350

[4] Speight JG. Refining heavy oil and extra-heavy oil. In: Delmon B, Yates JT, Centi G, editors. Heavy and Extra-Heavy Oil Upgrading Technologies. Vol. 164. New York: Elsevier; 2013. pp. 1-14

[5] Kilbane JJ. Microbial biocatalyst developments to upgrade fossil fuels. Current Opinion in Biotechnology. 2006;**17**:305-314

[6] Santos HF, Cury JC, do Carmo FL, et al. Mangrove bacterial diversity and the impact of petroleum contamination revealed by pyrosequencing: Bacterial proxies for petroleum pollution. PLoS One. 2011;**6**(3):e16943

[7] Ismail WA, van Hamme JD, Kilbane JJ, Gu J-D. Editorial: Petroleum microbial biotechnology: Challenges and prospects. Frontier in Microbiology. 2017;**8**:833

[8] Ehrlich HL. Geomicrobiology. New York: Marcel Dekker; 1995

[9] Epps AV. Phytoremediation of petroleum hydrocarbons. U.S. Environmental Protection Agency. 2006. Available from: http://clu-in.org/download/studentpapers/A_Van_EppsFinal.pdf [Accessed: December 2, 2011]

[10] Khan S, Afzal M, Iqbal S, Khan QM. Plant–bacteria partnerships for the remediation of hydrocarbon contaminated soils. Chemosphere. 2013;**90**(4):1317-1332

[11] Environment Agency (EA). Soil quality indicators. IPSS Meeting, Leeds. 2006

[12] Pinedo J, Ibáñez R, Lijzen JP, Irabien A. Assessment of soil pollution based on total petroleum hydrocarbons and individual oil substances. Journal of Environmental Management. 2013;**130**:72-79

[13] Rodríguez-Eugenio N, McLaughlin M, Pennock D. Soil Pollution: A Hidden Reality. Rome: FAO; 2018

[14] Ojuederie OB, Babalola OO. Microbial and plant-assisted bioremediation of heavy metal polluted environments: A review. International Journal of Environmental Resource and Public Health. 2017;**14**(12):1504

[15] Karthikeyan R, Kulakow PA. Soil plant microbe interactions in phytoremediation. In: Tsao DT, editor. Phytoremediation. Berlin, Heidelberg: Springer Berlin Heidelberg; 2003. pp. 52-74

[16] Ite AE, Ibok UJ. Role of plants and microbes in bioremediation of petroleum hydrocarbons contaminated soils. International Journal of Environmental Bioremediation and Biodgradation. 2019;**7**(1):1-19.s

[17] Obuekwe CO, Al-Muttawa EM. Self-immobilized bacterial cultures with potential for application as ready-to-use seeds for petroleum bioremediation. Biotechnology Letters. 2001;**23**: 1025-1032

[18] Kamath R, Rentz JA, Schnoor JL, Alvarez PJJ. Phytoremediation of Hydrocarbon-contaminated Soils: Principles and Applications. Iowa: Department of Civil and Environmental Engineering, Seamans Center, University of Iowa; 2007

[19] Shukla KP et al. Bioremediation: Developments, current practices and perspectives. Genetic Engineering and Biotechnology Journal. 2010;**2010**:GEBJ-3

[20] Coulon F, Brassington KJ, Bazin R, Linnet PE, Thomas KA, Mitchell TR, et al. Effect of fertilizer formulation and bioaugmentation on biodegradation and leaching of crude oils and refined products in soils. Environmental Technology. 2012;**33**(16):1879-1893

[21] Burken JG, Schnoor JL. Predictive relationships for uptake of organic contaminants by hybrid poplar trees. Environmental Science & Technology. 1998;**32**(21):3379-3385

[22] Siciliano SD, Fortin N, Mihoc A, Wisse G, Labelle S, Beaumier D, et al. Selection of specific endophytic bacterial genotypes by plants in response to soil contamination. Applied and Environmental Microbiology. 2001;**67**(6):2469-2475

[23] Lynch JM, Moffat AJ. Bioremediation—Prospects for the future application of innovative applied biological research. Annals of Applied Biology. 2005;**146**(2):217-221

[24] Chitara MJ, Chauhan S, Singh RP. Bioremediation of polluted soil by using plant growth-promoting rhizobacteria. In: Panpatte DG, Jhala YK, editors. Microbial Rejuvenation of Polluted Environment, Microorganisms for Sustainability. Springer Nature Singapore PVT Ltd; 2021. pp. 202-226

[25] Susarla S et al. Phytoremediation: An ecological solution to organic chemical contamination. Ecological Engineering. 2002;**18**(5):647-658

[26] U. S. Environmental Protection Agency (USEPA). A citizen's guide to phytoremediation. EPA-542-F-98-011. Washington DC: Office of Solid Waste and Emergency Response; 1998. 6 pp

[27] U.S. Department of Agriculture (USDA). Conservation buffers work—Economically and environmentally. Program Aid 1615. Washington, D.C.; 2000. 4 pp

[28] Ayotamuno J, Kogbara R, Agoro O. Biostimulation supplemented with phytoremediation in the reclamation of a petroleum contaminated soil. World Journal of Microbiology and Biotechnology. 2009;**25**(9):1567-1572

[29] Kumar R, Das A, Lal S. Petroleum hydrocarbon stress management in soil using microorganisms and their products. In: Chandra R, editor. Environmental Waste Management. Boca Raton: CRC Press; 2016. pp. 525-550

[30] Dickinson N. Phytoremediation. In: Murray BG, Murphy DJ, editors. Encyclopedia of Applied Plant Sciences. 2nd ed. Oxford: Academic Press; 2017. pp. 327-331

[31] Peer WA, Baxter IR, Richards EL, Freeman JL, Murphy AS. Phytoremediation and hyperaccumulator plants. In: Molecular Biology of Metal Homeostasis and Detoxification. Berlin: Springer; 2006. pp. 299-340

[32] Greipsson S. Phytoremediation. Nature Education Knowledge. 2011;**3**(10):7

[33] Jiao H, Luo J, Zhang Y, Xu S, Bai Z, Huang Z. Bioremediation of petroleum hydrocarbon contaminated soil by Rhodobacter sphaeroides biofertilizer and plants. Pakistan Journal of

Pharmaceutical Science. 2015;**28**(5 Suppl):1881-1886

[34] Al-Baldawi IA, Abdullah SRS, Anuar N, Suja F, Mushrifah I. Phytodegradation of total petroleum hydrocarbon (TPH) in diesel-contaminated water using Scirpus grossus. Ecological Engineering. 2015;**74**:463-473

[35] Hong MS, Farmayan WF, Dortch IJ, Chiang CY, McMillan SK, Schnoor JL. Phytoremediation of MTBE from a Groundwater Plume. Environmental Science & Technology. 2001;**35**(6):1231-1239

[36] U. S. Environmental Protection Agency (USEPA). Phytotechnologies for Site Cleanup. EPA 542-F-10-009. 2010

[37] Shiri M, Rabhi M, Abdelly C, Amrani AE. The halophytic model plant Thellungiella salsuginea exhibited increased tolerance to phenanthrene-induced stress in comparison with the glycophitic one Arabidopsis thaliana: Application for phytoremediation. Ecological Engineering. 2015;**74**:125-134

[38] Limmer M, Burken J. Phytovolatilization of organic contaminants. Environmental Science & Technology. 2016;**50**(13):6632-6643

[39] Masu S, Albulescu M, Balasescu L-C. Assessment on phytoremediation of crude oil polluted soils with Achillea millefolium and total petroleum hydrocarbons removal efficiency. Review Chimie. 2014;**65**:1103-1107

[40] Banks MK, Lee E, Schwab AP. Evaluation of dissipation mechanisms for Benzo[a]pyrene in the Rhizosphere of Tall Fescue. Journal of Environmental Quality. 1999;**28**(1):294-298

[41] Siciliano SD, Germida JJ. Enhanced phytoremediation of chlorobenzoates in rhizosphere soil. Soil Biology and Biochemistry. 1999;**31**(2):299-305

[42] Federal Remediation Technologies Roundtable (FRTR). Remediation Technologies Screening Matrix and Reference Guide, Version 4.0. 4.3 Phytoremediation (In situ Soil Remediation Technology). 2012. Available from: http://www.frtr.gov/matrix2/section4/4-3.html [Accessed: January 18, 2012]

[43] Pokethitiyook P. Phytoremediation of petroleum-contaminated soil in association with soil bacteria. In: Ansari AA, Gill SS, Gill R, Lanza GR, Newman L, editors. Phytoremediation: Management of Environmental Contaminants. Vol. 5. Cham: Springer International Publishing; 2017. pp. 77-99

[44] Marihal AK, Jagadeesh KS. Plant–microbe interaction: A potential tool for enhanced bioremediation. In: Arora NK, editor. Plant Microbe Symbiosis: Fundamentals and Advances. New Delhi: Springer India; 2013. pp. 395-410

[45] Marmiroli N, McCutcheon SC. Making phytoremediation a successful technology. In: McCutcheon SC, Schnoor JL, editors. Phytoremediation: Transformation and Control of Contaminants. John Wiley & Sons, Inc; 2004. pp. 85-119

[46] Andreolli M, Lampis S, Poli M, Gullner G, Biro B, Vallini G. Endophytic Burkholderia fungorum DBT1 can improve phytoremediation efficiency of polycyclic aromatic hydrocarbons. Chemosphere. 2013;**92**:688-694

[47] Schneider K, Oltmanns J, Radenberg T, Schneider T, Pauly-Mundegar D. Uptake of nitroaromatic compounds in plants. Environmental Science and Pollution Research. 1996;**3**(3):135-138

[48] Siciliano SD, Goldie H, Germida JJ. Enzymatic activity in root exudates of Dahurian Wild Rye (Elymus dauricus) that degrades 2-chlorobenzoic acid. Journal of Agricultural and Food Chemistry. 1998;**46**(1):5-7

[49] Alexander M. Biodegradation and Bioremediation. San Diego: Academic Press; 1994

[50] Anderson TA, Guthrie EA, Walton BT. Bioremediation in the rhizosphere. Environmental Science & Technology. 1993;**27**(13):2630-2636

[51] Salt DE, Smith RD, Raskin I. Phytoremediation. Annual Review of Plant Physiology and Plant Molecular Biology. 1998;**49**(1):643-668

[52] Phillips LA, Germida JJ, Farrell RE, Greer CW. Hydrocarbon degradation potential and activity of endophytic bacteria associated with prairie plants. Soil Biology and Biochemistry. 2008;**40**(12):3054-3064

[53] Moubasher HA, Hegazy AK, Mohamed NH, Moustafa YM, Kabiel HF, Hamad AA. Phytoremediation of soils polluted with crude petroleum oil using Bassia scoparia and its associated rhizosphere microorganisms. International Biodeterioration & Biodegradation. 2015;**98**(Supplement C):113-120

[54] Eapen S, D'Souza SF. Prospects of genetic engineering of plants for phytoremediation of toxic metals. Biotechnology Advances. 2005;**23**(2): 97-114

[55] Schwab AP, Banks MK. Biologically mediated dissipation of polyaromatic hydrocarbons in the root zone. In: Anderson TA, Coats JR, editors. Bioremediation through Rhizosphere Technology. Washington, DC (USA): American Chemical Society; 1994. pp. 132-141

[56] Cook RL, Hesterberg D. Comparison of trees and grasses for rhizoremediation of petroleum hydrocarbons. International Journal of Phytoremediation. 2013;**15**(9):844-860

[57] Njoku KL, Akinola MO, Oboh BO. Phytoremediation of crude oil contaminated soil: The effect of growth of Glycine max on the physico-chemistry and crude oil contents of soil. Nature and Science. 2009;**7**(10):79-87

[58] Merkl N, Schultze-Kraft R, Infante C. Assessment of tropical grasses and legumes for phytoremediation of petroleum-contaminated soils. Water, Air, and Soil Pollution. 2005;**165**(1):195-209

[59] Collins CD. Implementing phytoremediation of petroleum hydrocarbons. In: Willey N, editor. Phytoremediation: Methods and Reviews. Totowa, NJ: Humana Press; 2007. pp. 99-108

[60] Xia HP. Ecological rehabilitation and phytoremediation with four grasses in oil shale mined land. Chemosphere. 2004;**54**(3):345-353

[61] Khan AG. Role of Vetiver Grass and Arbuscular Mycorrhizal fungi in improving crops against abiotic stresses. In: Ashraf M, Ozturk M, Athar HR, editors. Salinity and Water Stress: Improving Crop Efficiency. Dordrecht: Springer Netherlands; 2009. pp. 111-116

[62] Lu M, Zhang Z, Sun S, Wei X, Wang Q, Su Y. The use of Goosegrass (Eleusine indica) to remediate soil contaminated with petroleum. Water, Air, & Soil Pollution. 2010;**209**(1): 181-189

[63] Ayotamuno JM, Kogbara RB, Egwuenum PN. Comparison of corn and elephant grass in the phytoremediation of a petroleum-hydrocarbon-contaminated agricultural soil in Port Harcourt, Nigeria. Journal of Food, Agriculture & Environment. 2006;**4**(3,4):1-12

[64] Meeinkuirt W, Kruatrachue M, Tanhan P, Chaiyarat R, Pokethitiyook P. Phytostabilization potential of Pb mine tailings by two grass species, Thysanolaena maxima and Vetiveria zizanioides. Water, Air, & Soil Pollution. 2013;**224**(10):1750

[65] Frick CM, Germida JJ, Farrell RE. Assessment of phytoremediation as an in-situ technique for cleaning oil-contaminated sites. Technical Seminar on Chemical Spills. 1999;**16**:105a-124a

[66] Cherian S, Oliveira MM. Transgenic plants in phytoremediation: Recent advances and new possibilities. Environmental Science & Technology. 2005;**39**(24):9377-9390

[67] Agency for Toxic Substances and Disease Registry (ATSDR). Toxicological profile for total petroleum hydrocarbons (PHC). Atlanta, GA: U.S. Department of Health and Human Services, Public Health Service; 1999. Available from: http://www.atsdr.cdc.gov/ToxProfiles/tp123.pdf [Accessed: March 20, 2012]

[68] Margesin R, Schinner F. Biodegradation and bioremediation of hydrocarbons in extreme environments. Applied Microbiology and Biotechnology. 2001;**56**:650-663

[69] MERCK. "Indigenous Plants to the Rescue." Science in Africa. 2002. Available from: http://www.scienceinafrica.co.za/2002/february/oil.htm [Accessed: January 15, 2012]

[70] Chiapusio G et al. Phenanthrene toxicity and dissipation in rhizosphere of grassland plants (Lolium perenne L. and Trifolium pratense L.) in three spiked soils. Plant Soil. 2007;**294**:103-112

[71] Wenzel WW. Rhizosphere processes and management in plant-assisted bioremediation (phytoremediation) of soils. Plant Soil. 2009;**321**:385-408

[72] Bian X-Y, Mbadinga SM, Liu Y-F, Yang S-Z, Liu J-F, Ye R-Q, et al. Insights into the anaerobic biodegradation pathway of n-alkanes in oil reservoirs by detection of signature metabolites. Scientific Reports. 2015;**5**:9801

[73] Chaineau CH, Yepremian C, Vidalie JF, Ducreux J, Ballerini D. Bioremediation of a crude oil-polluted soil: Biodegradation, leaching and toxicity assessment. Water, Air, and Soil Pollution. 2003;**144**:419-440

[74] Bachmann R, Johnson A, Edyvean R. Biotechnology in the petroleum industry: An overview. International Biodeterioration and Biodegradation. 2014;**86**:225-237

[75] Miller AJ, Cramer MD. Root nitrogen acquisition and assimilation. Plant and Soil. 2004;**274**:1-36

[76] Suominen L, Jussila MM, Mäkeläinen K, Romantschuk M, Lindström K. Evaluation of the Galega-Rhizobium galegae system for the bioremediation of oilcontaminated soil. Environmental Pollution. 2000;**107**: 239-244

[77] Havlin JL, Beaton JD, Tisdale SL, Nelson WL. Soil Fertility and Fertilizers: An Introduction to Nutrient Management. 7th ed. Published by Pearson Education, Inc; 2010

[78] Yadegari M, Rahmani HA, Noormohammadi G, Ayneband A. Plant growth promoting rhizobacteria increase growth, yield and nitrogen fixation in Phaseolus vulgaris. Journal of Plant Nutrition. 2010;**33**(12):1733-1743

[79] CLUIN phytoremediation database. Available from: http://cluin.org/products/phyto/ [Accessed: August 20, 2012]

[80] Interstate Technology & Regulatory Council (ITRC). Phytotechnology Technical and Regulatory Guidance and Decision Trees, Revised. Phyto-3. Washington, DC (USA). 2009. Available from: http://www.itrcweb.org/Documents/PHYTO-3.pdf [Accessed: December 29, 2011]

[81] Peng S, Zhou Q, Cai Z, Zhang Z. Phytoremediation of petroleum contaminated soils by Mirabilis jalapa L. in a greenhouse plot experiment.

Journal of Hazardous Materials. 2009;**168**(2):1490-1496

[82] Zhang Z, Rengel Z, Chang H, Meney K, Pantelic L, Tomanovic R. Phytoremediation potential of Juncus subsecundus in soils contaminated with cadmium and polynuclear aromatic hydrocarbons (PAHs). Geoderma. 2012;**175**:1-8

[83] Afzal M, Yousaf S, Reichenauer TG, Kuffner M, Sessitsch A. Soil type affects plant colonization, activity and catabolic gene expression of inoculated bacterial strains during phytoremediation of diesel. Journal of Hazardous Materials. 2011;**186**(2):1568-1575

[84] Gkorezis P, Daghio M, Franzetti A, Van Hamme JD, Sillen W, Vangronsveld J. The interaction between plants and bacteria in the remediation of petroleum hydrocarbons: An environmental perspective. Frontiers in Microbiology. 2016;7:1836

[85] Spada V, Iavazzo P, Sciarrillo R, Guarino C. Successful integrated bioremediation system of hydrocarbon-contaminated soil at a former oil refinery using autochthonous bacteria and rhizo-microbiota. In: Ansari AA, Gill SS, Gill R, Lanza GR, Newman L, editors. Phytoremediation: Management of Environmental Contaminants. Vol. Volume 5. Cham: Springer International Publishing; 2017. pp. 53-76

[86] Tewari S, Arora NK. Transactions among microorganisms and plant in the composite rhizosphere habitat. In: Arora NK, editor. Plant Microbe Symbiosis: Fundamentals and Advances. New Delhi: Springer India; 2013. pp. 1-50

[87] Wojtera-Kwiczor J, Żukowska W, Graj W, Małecka A, Piechalak A, Ciszewska L, et al. Rhizoremediation of diesel-contaminated soil with two rapeseed varieties and petroleum degraders reveals different responses of the plant defense mechanisms. International Journal of Phytoremediation. 2014;**16**(7-8):770-789

[88] Erickson LE. An overview of research on the beneficial effects of vegetation in contaminated soil. Annals of the New York Academy of Sciences. 1997;**829**(1):30-35

[89] Singh RP, Dhania G, Sharma A, Jaiwal PK. Biotechnological approaches to improve phytoremediation efficiency for environment contaminants. In: Singh SN, Tripathi RD, editors. Environmental Bioremediation Technologies. Berlin, Heidelberg: Springer Berlin Heidelberg; 2007. pp. 223-258

[90] Kuiper I, Lagendijk EL, Bloemberg GV, Lugtenberg BJJ. Rhizoremediation: A beneficial plant-microbe interaction. Molecular Plant-Microbe Interactions. 2004;**17**(1):6-15

[91] Lugtenberg BJJ, Kravchenko LV, Simons M. Tomato seed and root exudate sugars: Composition, utilization by Pseudomonas biocontrol strains and role in rhizosphere colonization. Environmental Microbiology. 1999;**1**(5):439-446

[92] Meng L, Qiao M, Arp HPH. Phytoremediation efficiency of a PAH-contaminated industrial soil using ryegrass, white clover, and celery as mono- and mixed cultures. Journal of Soils and Sediments. 2011;**11**(3): 482-490

[93] Ajuzieogu CA, Ibiene AA, Stanley HO. Laboratory study on influence of plant growth promoting rhizobacteria (PGPR) on growth response and tolerance of Zea mays to petroleum hydrocarbon. African Journal of Biotechnology. 2015;**14**(43):2949-2956

[94] Asad SA. Soil–PCB–PGPR interactions in changing climate scenarios. In: Xenobiotics in the Soil

Environment. Cham: Springer; 2017. pp. 281-298

[95] Ma Y, Prasad MN, Rajkumar M, Freitas H. Plant growth promoting rhizobacteria and endophytes accelerate phytoremediation of metalliferous soils. Biotechnology Advances. 2011;**29**: 248-258

[96] Bhattacharyya PN, Jha DK. Plant growth-promoting rhizobacteria (PGPR): Emergence in agriculture. World Journal of Microbiology and Biotechnology. 2012;**28**(4):1327-1350

[97] Cassán F, Vanderleyden J, Spaepen S. Physiological and agronomical aspects of phytohormone production by model plant-growth-promoting rhizobacteria (PGPR) belonging to the genus Azospirillum. Journal of Plant Growth Regulators. 2014;**33**(2):440-459

[98] Podile AR, Vukanti RV, Sravani A, Kalam S, Dutta S, Durgeshwar P, et al. Root colonization and quorum sensing are the driving forces of plant growth promoting rhizobacteria (PGPR) for growth promotion. Proceedings of the National Academic Science. 2014;**80**(2):407-413

[99] Nadeem SM, Ahmad M, Zahir ZA, Javaid A, Ashraf M. The role of mycorrhizae and plant growth promoting rhizobacteria (PGPR) in improving crop productivity under stressful environments. Biotechnology Advances. 2014;**32**:429-448

[100] Patel PR, Shaikh SS, Sayyed RZ. Dynamism of PGPR in bioremediation and plant growth promotion in heavy metal contaminated soil. Indian Journal of Experimental Biology. 2016;**54**:286

[101] Antony R, Sujith PP, Sheryl OF, Pankaj V, Khedekar VD, Loka Bharathi PA. Cobalt immobilization by manganese oxidizing bacteria from Indian Ridge System. Current Microbiology. 2011;**62**:840-849

[102] Cheung KH, Gu JD. Mechanism of hexavalent chromium detoxification by microorganisms and bioremediation application potential: A review. International Biodeterioration and Biodegradation. 2007;**59**:8-15

[103] Bacosa HP, Inoue C. Polycyclicaro matichydrocarbons(PAHs) biodegradationpotentialand diversity of microbial consortia enriched from tsunami sediments in Miyagi, Japan. Journal of Hazardous Mater. 2015;**283**:689-697

[104] Kotoky R, Das S, Singha LP, Pandey P, Singha KM. Biodegradation of Benzo (a) pyrene by biofilm forming and plant growth promoting Acinetobacter sp. strain PDB4. Environmental Technology Innovation. 2017;**8**:256-268

[105] Chen W, Li J, Sun X, Min J, Hu X. High efficiency degradation of alkanes and crude oil by a salt-tolerant bacterium Dietzia species CN-3. International Biodeterioration and Biodegradation. 2017;**118**:110-118. DOI: 10.1016/j.ibiod.2017.01.029

[106] Gangola S, Kumar R, Sharma A, Singh H. Bioremediation of petrol engine oil polluted soil using microbial consortium and wheat crop. Journal of Pure Applied Microbiology. 2017;**11**(3):1583-1588

[107] Bhakat K, Chakraborty A, Islam E. Characterization of arsenic oxidation and uranium bioremediation potential of arsenic resistant bacteria isolated from uranium ore. Environment Science and Pollutant Research. 2019;**26**(13):12907-12919

[108] Hou J, Liu W, Wang B, Wang Q, Luo Y, Franks AE. PGPR enhanced phytoremediation of petroleum contaminated soil and rhizosphere microbial community response. Chemosphere. 2015;**138**:592-598

[109] Liu JL, Xie BM, Shi XH, Ma JM, Guo CH. Effects of two plant

growth-promoting rhizobacteria containing 1-aminocyclopropane-1-carboxylate deaminase on oat growth in petroleum-contaminated soil. International Journal of Environmental Science and Technology. 2015;**12**(12):3887-3894

[110] Hazotte A, Péron O, Gaudin P, Abdelouas A, Lebeau T. Effect of Pseudomonas fluorescens and pyoverdine on the phytoextraction of cesium by red clover in soil pots and hydroponics. Environment Science and Pollutant Research. 2018;**25**(21): 20680-20690

[111] Compant S, Clément C, Sessitsch A. Plant growth-promoting bacteria in the rhizo-and endosphere of plants: Their role, colonization, mechanisms involved and prospects for utilization. Soil Biology and Biochemistry. 2010;**42**(5):669-678

[112] Gao X, Gao W, Cui Z, Han B, Yang P, Sun C, et al. Biodiversity and degradation potential of oil-degrading bacteria isolated from deep-sea sediments of South MidAtlantic Ridge. Marine Pollution Bulletin. 2015;**97**:373-380

[113] Peixoto RS, Vermelho AB, Rosado AS. Petroleum-degrading enzymes: Bioremediation and new prospects. Enzyme Research. 2011;**34**:1-8

[114] Broderick JB. Catechol dioxygenases. Essays in Biochemistry. 1999;**34**(11):173-189

[115] Cao B, Nagarajan K, Loh KC. Biodegradation of aromatic compounds: Current status and opportunities for biomolecular approaches. Applied Microbiology and Biotechnology. 2009;**85**(2):207-228

[116] Madigan MT, Martinko JM, Dunlap PV, Clark DP. Brock Biology of Microorganisms. 12th ed. Benjamin Cummings; 2010

[117] Whyte LG, Bourbonniere L, Greer CW. Biodegradation of petroleum hydrocarbons by psychotrophic Pseudomonas strains possessing both alkane (alk) and napthalene (nah) catabolic pathways. Applied Environmental Microbiology. 1997;**63**:3719-3723

[118] Steliga T, Jakubowicz P, Kapusta P. Changes in toxicity during in situ bioremediation of weathered drill wastes contaminated with petroleum hydrocarbons. Bioresources Technology. 2012;**125**:1-10

[119] Stosky JB, Greer CW, Atlas RM. Frequency of genes in aromatic and aliphatic hydrocarbon biodegradation pathways within bacterial populations from Alaskan sediments. Canadian Journal of Microbiology. 1994;**40**: 981-985

[120] van Beilen JB, Funhoff EG. Alkane hydroxylases involved in microbial alkane degradation. Applied Microbiology and Biotechnology. 2007;**74**(1):13-21

[121] van Beilen JB, Wubbolts MG, Witholt B. Genetics of alkane oxidation by Pseudomonas oleovorans. Biodegradation. 1994;**5**(3-4):161-174

[122] Kweon O, KimSJ BJ, Kim SK, Kim BS, Baek DH, Park SI, et al. Comperative functional pan-genome analyses to build connections between genomic dynamics and phenotypic evolution in polycyclic aromatic hydrocarbon metabolism in the genus Mycobacterium. BMC Evolutionary Biology. 2015;**15**:21

[123] Zhou J, Bian X-Y, Mbadinga SM, Yang S-Z, Yang J-F, Gu J-D, et al. Synthesis and characterization of anaerobic degradation biomarkers of n-alkanes via hydroxylation/ carboxylation pathways. European Journal of Mass Spectroscopy. 2016;**22**:31-37. DOI: 10.1255/ejms.1402

[124] van Beilen JB, Panke S, Lucchini S, Franchini AG, Rothlisberger M, Witholt B. Analysis of Pseudomonas putida alkane-degradation gene clusters and flanking insertion sequences: Evolution and regulation of the alk genes. Microbiology. 2001;**147**(6): 1621-1630

[125] Rabus R, Boll M, Heider J, Meckenstock RU, Buckel W, Einsle O, et al. Anaerobic microbial degradation of hydrocarbons: From enzymatic reactions to the environment. Journal of Molecular Microbiology and Biotechnology. 2016;**26**:5-28

[126] Alcalde M, Ferrer M, Plou FJ, Ballesteros A. Environmental biocatalysis: From remediation with enzymes to novel green processes. Trends in Biotechnology. 2006;**24**(6): 281-287

[127] Sutherland TD, Horne I, Weir KM, et al. Enzymatic bioremediation: From enzyme discovery to applications. Clinical and Experimental Pharmacology and Physiology. 2004;**31**(11):817-821

[128] Setti L, Lanzarini G, Pifferi PG. Whole cell biocatalysis for an petroleum desulfurization process. Fuel Processing Technology. 1997;**52**(1-3):145-153

[129] Alcalde M, Bulter T, Zumarraga M, et al. Screening mutant libraries of fungal laccases in the presence of organic solvents. Journal of Biomolecular Screening. 2005;**10**(6): 624-631

[130] Smith M, Thurston F, Wood DA. Fungal laccases: Role in delignification and possible industrial applications. In: Messerschmidt A, editor. Multi-Copper Oxi-Dases. Singapore: World Scientific Publishing; 1997. pp. 201-224

[131] Bulter T, Alcalde M, Sieber V, Meinhold P, Schlachtbauer C, Arnold FH. Functional expression of a fungal laccase in Saccharomyces cerevisiae by direct edevolution. Applied and Environmental Microbiology. 2003;**69**(2):987-995

[132] Scott C, Lewis SE, Milla R, et al. A free-enzyme catalyst for the bioremediation of environmental atrazine contamination. Journal of Environmental Management. 2010;**91**(10):2075-2078

[133] Whiteley CG, Lee DJ. Enzyme technology and biological remediation. Enzyme and Microbial Technology. 2006;**38**(3-4):291-316

[134] U. S. Environmental Protection Agency (USEPA). Oil: Crude and Petroleum Products. 2011. Available from: http://www.eia.gov/energyexplained/index.cfm?page=oil_home [Accessed: December 24, 2011]

Chapter 4

# Potential Defensive Involvement of Methyl Jasmonate in Oxidative Stress and Its Related Molecular Mechanisms

*Gunjegaonkar Shivshankar M., Joshi Amol A., Wankhede Sagar B, Siraskar Balasaheb D., Merekar Abhijit N. and Shinde Sachin D.*

## Abstract

Jasmonic acid (JA), cytokinins (CK), gibberellins (GA), abscisic acid (ABA), ethylene (ET), and salicylic acid (SA) are potent plant stress hormones (phytohormones/PTH). Methyl jasmonate (MeJA), a volatile ester of JA, is derived from the petals of Jasminum grandiflorum (jasmine). The MeJA has been meticulously confirmed for its food, agricultural, and therapeutic uses in the treatment of a range of serious illnesses. Several scientific articles have studied and reported on the role of free radicals in the development of life-threatening clinical illnesses. The inflammatory signaling pathway is triggered by a weak or interfering endogenous antioxidant system, or the elaborated production of free radicals, which causes damage to key cellular components. The current chapter focused on and demonstrated MeJA's multifunctional role in antioxidant and anti-inflammatory signaling mechanisms such as inhibition of NF-B (nuclear factor kappa-light-chain-enhancer of activated B cells), mitogen-activated protein kinase (MAPK or MAP kinase) pathway inhibition/down-regulation of pro-inflammatory mediators (IL, TNF-), cyclo-oxygenase (COX), and (LOX). The antioxidant effect of MeJA's interaction with miRNA, transcription of nuclear factor erythroid 2-related 2 (Nfr2), activation of sirtuins (SIRTs), antioxidant and redox signaling pathway were also discussed in the chapter.

**Keywords:** methyl jasmonate, anti-inflammatory, oxidative stress, free radicals, plant stress hormone

## 1. Introduction

Oxygen is continually used by the organism for a variety of important activities [1]. Reactive oxygen species (ROS) is a naturally occurring byproduct of oxygen metabolism that interacts with biological systems regularly [2, 3]. ROS has been involved in the breakdown of cell organelles such as DNA, proteins, and lipids, according to several studies. Radical scavenging molecules are important components of the antioxidative defense mechanism, protecting cells from free radical damage [4, 5]. Endogenous antioxidant systems (enzymatic/nonenzymatic) are important for balancing and fighting ROS such as $H_2O_2$, ROOR (organic hydroperoxide), NO (nitric oxide), O-(superoxide), and •OH (hydroxyl radicals), among others [6, 7]. Superoxide dismutase (SOD), catalase (CAT), glutathione (GSH), and glutathione reductase (GR) are enzymes in the human endogenous antioxidant system that play a role in the development of ROS-mediated oxidative stress [7–10]. The extensive production of free radicals manifests in serious illnesses and disorders as diabetes [11], cardiovascular [12], inflammatory [13], and neurological diseases [14]. The production of ROS is triggered by many oxidizing enzymes (xanthine oxidases, NADPH oxidase) COX [15, 16] pro-inflammatory factors. A modest rise in ROS levels disturbs and interferes with cell proliferation and normal physiological processes, but a substantial increase in ROS levels causes catastrophic damage to cellular components [17, 18]. SOD, an important endogenous antioxidant metalloenzyme that scavenges $H_2O_2$ produced during oxygen metabolism [19], is an important endogenous antioxidant metalloenzyme. Because of the increased and unregulated synthesis of O-depletes in SOD storage, protecting the cell from hazardous chemicals created during aerobic respiration becomes challenging [20]. Through the conversion of GSH into selenium-containing GSH (GSSH) by GR (Glutathione reductase), the thiol-containing GSH is an essential reducing agent that eliminates reactive oxygen species [20, 21]. CATase is a kind of antioxidant that is responsible for the neutralization of $H_2O_2$. $H_2O_2$ causes lipid peroxidation, which is inhibited by CAT [22]. Many compounds are produced by lipid peroxidation, but one of the most significant is MDA (malondialdehyde) [9].

## 2. Multifunctional role of MeJA in several clinical ailments

### 2.1 Downregulation of NF-κB, MAPK pathway, and oxidative stress

The cytokines are important mediators and signaling molecules involved in several inflammatory pathological cascades. TNF- and the IL family (IL-1, IL-6, IL-15, IL-17, IL-18) are important cytokines [23]. The vital regulator in the synthesis of cytokine induction is the NF-κB signaling pathway. NF-κB is a family of proteins associated with DNA and activates cytokine synthesis genes [24]. The inhibitory protein (IkB) binds to NF-κB and blocks the translocation it from the cytoplasm to the nucleus. The number of stimuli affects IkB function by phosphorylation and degrading it. This allows NF-B to get activated, translocated to the nucleus from the cytoplasm, and triggers genes involved in the synthesis of inflammatory mediators. Agents that activate NF-κB include cytokines, mitogens, and ROS [25]. Another important signaling pathway in inflammation is the activation of MAP kinase under the influence of ROS. ROS induces phosphorylation of the MAP kinase family, including ERK, JNK, and p38 kinases [26]. Consequently, these kinases are responsible for the activation of transcription factors similar to NF-κB and regulate pro-inflammatory genes. Through enzymatic conversion from LOX and COX, ROS also significantly contributed to the synthesis of inflammatory mediators such as PG (prostaglandins), LT (leukotrienes), 5-HETE (5-Hydroxyeicosatetraenoic acid), and others [27, 28]. Studies have shown that antioxidant treatment effectively reduces pain and inflammation via downregulation of NF-B activation and cytokine synthesis [29, 30]. Nociceptive responses are detected by nociceptors found on c and fibers that innervate higher brain centers. Several studies have advocated that the activation of nociceptors and perception is mediated by inflammatory mediators [31]. The study revealed that MeJA significantly reduces the production of ROS-mediated oxidative stress and the generation of pro-inflammatory mediators and would be responsible for its anti-inflammatory and anti-nociceptive effects in-vivo and in-vitro [28, 32–35]. Studies have demonstrated a significant reduction in pain and inflammation in MeJA-treated experimental animals intoxicated with LPS. The proposed underlying mechanism involved is downregulating the production of pro-inflammatory cytokines (IL, TNF-), expression of COX and LOX, PG, resolving disturbed redox status, inhibiting the generation of ROS/RNS, inactivation of inflammatory cells, and downregulation of transcription in the NF-B and MAPK signaling pathways [35].

### 2.2 Inhibition of neuronal peroxidation and oxidative stress

The progressive loss of memory is characterized by Alzheimer's disease (AD). The common signs and symptoms include the inability to recall past events, calculate, plan and perform simple tasks, recognize people and relationships, etc. [36, 37]. The pathological lesion lies in central cholinergic pathways where the degeneration of cholinergic neuronal populations occurs [38]. Several studies have revealed that ROS and associated oxidative stress in the brain region is important in the development and progression of AD. Enhanced lipid peroxidation and diminished polyunsaturated fatty acids have been found in AD brains, which further support the role of ROS in the pathophysiology of the disease [39, 40]. Brain tissues are more susceptible to the deleterious effects of ROS because of their high rate of oxygen consumption, high iron content in many brain tissues, and generation of hydrogen peroxide in neuronal mitochondria cells [41, 42]. Postmortem studies have confirmed elevated levels of MDA, an index of lipid peroxidation in AD brains, which confirmed the role of oxidative stress in the pathogenesis of the disease [43]. Induction of AD in experimental animals can be done by the administration of chemical substances which interfere in a central cholinergic pathway. Scopolamine (SC)-induced memory dysfunction has been linked to its depletion of Ach (acetylcholine) stores, increased oxidative stress, and depletion of the endogenous antioxidant enzymes in brain tissues, which leads to neuronal damage. Tacrine, an AChE inhibitor like tacrine, was the first further donepezil and rivastigmine to be approved for the treatment of AD. It is reported that inflammation of brain cells appears to contribute to the development and progression of AD. Anti-inflammatory drugs such as NSAIDs, corticosteroids, and antioxidants may be effective strategies in Alzheimer's disease [42, 43]. In a pharmacological screening of a new molecule, lipopolysaccharide (LPS) induced neuroinflammation is a highly validated and reported model [44, 45]. The underlying mechanism involves LPS-induced synthesis of inflammatory mediators and generation of ROS followed by damage to the neurons. MeJA has been screened for its memory performance-enhancing potential against LPS-induced neurotoxicity. The studies revealed that significant memory enhancement in MeJA treated animals was observed as compared to LPS treated. The underlying mechanism has been linked to MeJA antioxidant, anti-inflammatory potential using In-vivo as well as In-Vitro screening models [46–48].

### 2.3 Inhibition of mitochondrial dysfunctioning, inflammatory cytokines, and oxidative stress

Stress can influence the emotional factors and neurobehavioral characteristics of human beings and manifest anxiety [49]. The association of oxidative stress and neurodegenerative disorders, including anxiety, has been reported in several studies [50]. Anxiety is a stress response such as worry, fear, overwhelm, and distress to the environment that makes it difficult to continue to work or behave normally in day-to-day life [51]. The stress system components of the CNS are the limbic system, hypothalamus, pituitary, and endocrine hormones that play an integral part in the determination of mental health and behavioral responses [52]. These behavioral responses are regulated by the neurotransmitter/modulators and get interrupted by a variety of chemicals, xenobiotics, drugs, etc., and could change the normal neuronal function [53]. The brain tissues are rich in lipid substrates for oxidation, iron, and copper ions that catalyze free radical reactions which are abundant [54]. The sites of damage mediated by ROS are neuronal mitochondria dysfunction, which leads to psychiatric behavioral diseases like depression, anxiety, psychosis, and ataxia [55, 56]. Patients with anxiety, depression, and psychosis found enhanced levels of pro-inflammatory cytokines, specifically IL-6 and TNF-α, in their blood and brain. Pro-inflammatory cytokines degenerate neurons by activating signaling molecules such as phospholipase A2 (PLA2) and arachidonic acid (AA) [57, 58]. Activation of PLA2 and AA further increases ROS and additional inflammatory mediators like eicosanoids, which contribute to promoting inflammation and nerve degeneration [59]. AA has been found to have a direct role in apoptotic effects [63]. Anti-inflammatory, antioxidant dietary agents such as docosahexaenoic acid (DHA) have been shown in studies to prevent neuronal apoptosis and to be an important treatment option in neurodegenerative diseases [61]. NF-B expression has been found in neurodegenerative diseases, including anxiety [60]. Blocking of the NF-κB pathway remarkably reduces the levels of cytosolic and mitochondrial ROS generation and neuronal damage mediated by oxidative stress. MeJA has been studied for its antianxiety, adaptogenic, and anti-stress potential, and it has been shown to have a significant effect in animals [49, 61]. The underlying mechanism is that MeJA significantly reduces the levels of mitochondrial ROS by compensating with endogenous antioxidant enzymes like GSH, CAT, GPx, GR, SOD, and free radical scavenging activity. A significant reduction in IL and TNF-, as well as a significant inhibition of NO, which is responsible for the synthesis of pro-inflammatory mediators, has a direct effect on the inhibition of neurodegeneration. It was also reported that MeJA shows a switch-off effect on activation of the NF-B and MAPK transcription pathways, which have direct involvement in the generation of stress, anxiety, and other psychological disturbances [31, 34].

### 2.4 Inhibition of neuronal excitability and oxidative stress

Aggressive tendencies and behaviors in humans and animals have been demonstrated and linked to elevated inflammatory markers [62, 63]. As discussed earlier, brain tissues are more susceptible to the deleterious effects of ROS because of their high rate of oxygen consumption, high iron content in many brain tissues, and generation of hydrogen peroxide in neuronal mitochondria cells [40, 41]. ROS damages neuronal membranes, impairs the ability to deactivate receptors and ion channels, causes uncontrolled neurotransmitter release, and generally disrupts neuronal functioning [64, 65]. disturbed neuronal functioning contributes to neuronal excitotoxicity and is considered as a pathological cascade in neuronal diseases, particularly aggression. Excess excitatory neurotransmitter (glutamate) activity and weakened inhibitory (GABA) neurotransmitter signaling result in aggressive behavior changes. Over-activity of glutamate excitatory neurotransmitters progressively modulates the glutamate receptor and increases intracellular levels of Ca2+, which further disturbs the mitochondrial Ca2+ homeostasis, activates hydrolytic enzymes, and activates apoptotic signaling pathways. Several studies have established a link between increased nitric oxide synthase (NOS) activity and glutamate neurotoxicity, as well as associated behavioral aggressive tendencies [66, 67]. Evidence suggests that MeJA treatment can control disease progression by modulating oxidative stress-mediated by ROS/RNS and controlling inflammatory mediators via direct or indirect mechanisms [61, 68]. In conclusion, the antioxidant potential of MeJA controls the modulation of glutamate/GABAergic, increased Ca2+ influx through countering NO production, oxidative stress.

### 2.5 Inhibition of metalloproteinases and oxidative stress

The articular cartilage is avascular and does not receive any blood supply. Hence, the essential nutrients and oxygen are supplied to the cartilage through the synovial fluid [69]. Many metabolic reactions in chondrocytes are anaerobic and adapted to survive with a minimum oxygen tension [70]. In a pathological condition, oxygen tension fluctuates, leading to the generation of ROS by the chondrocytes. The main reactive species produced by chondrocytes are $O_2$ radicals, NO, and their derivatives (ONOO-, $H_2O_2$) example, chondrocyte-derived free radical levels are important for the maintenance of ion homeostasis, but they also contribute to disease progression [71]. Enhanced levels of ROS lead to serious damage to both chondrocytes and extracellular matrix components of articular cartilage and disturb redox status [72]. The important component of the ECM like aggrecan is degraded by ONOO-and initiates the process of cartilage degradation [73]. Additionally, endogenously synthesized NO suppresses the synthesis of aggrecan. The tensile strength is primarily provided by a network of aggrecan hyaluronate collagen; free radicals disturb the collagen network and reduce the strength of the ECM. Free radicals inhibit collagen synthesis indirectly via interleukin-1 [74]. The proteoglycan synthesis is inhibited by $H_2O_2$ through the disturbing synthesis of triphosphate (ATP) [75]. The tissue inhibitors of metalloproteinases (TIMPs) are important inhibitors of MMP-mediated cartilage damage [76]. ONOO-and HOCl reduces the activity of TIMPs by inactivating them. The NO producing agent up regulated the synthesis of MMPs by enhancing collagenase mRNA expression [80]. Proteoglycan synthesis is down-regulated in the chondrocytes on exposure to $H_2O_2$ [77]. ROS participates in reducing the capacity of chondrogenic precursor cells to migrate and proliferate within the joint area. NO enhances the anti-proliferative effect of IL-1 as well as initiates chondrocyte apoptosis [78]. The fibroblast-like synoviocytes consume a large amount of oxygen as compared to chondrocytes. In oxidative stress, the accumulation of antioxidant enzymes like SOD, CAT, and GSH has been observed. These enzymes protect the chondrocytes and ECM degradation from free radicals [79]. An uncontrolled and abnormal increase in ROS levels causes apoptosis of chondrocytes. Several studies reveal that a minimum level of $H_2O_2$ in synoviocytes causes less damage to chondrocytes [80]. MeJA has shown a significant chondroprotective effect on LPS-induced cartilage damage. LPS induces the synthesis of pro-inflammatory mediators as well as creates severe oxidative stress. MeJA significantly reduces the inflammatory mediator's activity as well as cartilage destructive MMP. Normalization of oxidative stress is accomplished by restoring antioxidant enzyme levels [27, 32].

## 3. Regulation of miRNA, SIRT, and HIF1α for an antioxidant mechanism

An oxidative stress state alters the expression level of different miRNAs (microRNAs) and causes significant changes in important cellular processes like cell differentiation, lipid metabolism, apoptosis, and organ development [81]. Severe clinical conditions like inflammation, cancer, cardiovascular diseases, diabetes mellitus, rheumatoid arthritis, neurological disorders have been correlated with altered miRNA expression [82]. Upregulation or downregulation of mRNA addresses pathophysiological modulation in retardation or development of diseases. Anticancer activity of MeJA against bladder, colorectal cancer cells has been shown via the downregulation of EZH2 (enhancer of zeste homolog 2) expression by induction of microRNA-101 [83–85]. MeJA antioxidant activity has been demonstrated via two pathways: first, inhibition or down-regulation of pro-inflammatory factors such as IL, TNF-mediated mitochondrial ROS production, and second, restoration of endogenous antioxidant enzymes [86, 87]. The latter mechanism involves an effect on the regulation of microRNA. Cellular redox status is regulated by redox-sensible transcription Nfr2 through the upregulation of antioxidant defense genes for SOD, CAT, GSH enzymes [88–90]. The activity of Nfr2 is regulated by miRNAs via downregulating the same [91, 92]. In conclusion, MeJA improves the antioxidant status and regulates oxidative stress by downregulating specific miR-155 and upregulating miR-101, which leads to the upregulation of Nfr2 activity. The indirect effect of MeJA on sirtuins (SIRTs) as an antioxidant and redox signaling pathway has been established through upregulation of Nfr2 activity by induction of miR-101. Researchers have extensively studied and reported that SIRTs are key signaling molecules that regulate the redox status of the cell and modulate cellular responses in a variety of pathological conditions over the last two decades [93]. SIRTs protect the cell from the deleterious effect of ROS and enhance the expression of genes responsible for the production of endogenous antioxidant enzymes. SIRTs are important for the fine balance between oxidant and antioxidant systems, regulating cellular biochemical reactions as well as maintaining an oxidative state. SIRTs, in association with antioxidant response elements (ARE), is involved in the regulation of gene expression when a cell is exposed to oxidative stress responses. ARE senses the altered cellular redox status and elicits transcriptional responses through activation of Nfr2. Nfr2 regulates the expression and production of several antioxidant enzymes and detoxification genes [94, 95]. Several studies have reported that MeJA significantly restores the levels of antioxidant enzymes and reduces oxidative stress and mediated damage [27, 50, 51, 96, 97]. The underlying mechanism may be Nfr2 mediated increased gene transcription for antioxidant enzymes through the upregulation of miRNA-101. MeJA down-regulates the expression of miRNA-155, leading to the stabilization of HIF-1 (hypoxia-inducible factor 1 alpha) [88]. The fall in oxygen tension in the cell below that needed for normal physiological demand causes a cellular hypoxic adaptive response. The hypoxic condition is crucial and important to target for a therapeutic approach, particularly in cardiovascular disease and cancer [98]. The key regulators of oxygen tension in cells are HIFs. HIF-1 regulates acute hypoxia, whereas HIF-2 and HIF-3 regulate chronic hypoxia. Recently, investigation suggests that miRNAs play an important role in the regulation of HIF [99]. Chronic hypoxic conditions lead to cellular apoptosis, which contributes to severe stroke or myocardial infarction. Alternatively, an intentional cellular hypoxia approach is practiced in the treatment of various types of cancer. Hypoxia induces oxidative stress via the overgeneration of reactive oxygen species (ROS). Targeting HIF through downregulation of miRNA-155 is a new dimension in the induction of cancer cell apoptosis [100]. MeJA anticancer activity has been largely correlated with the downregulation of miRNA-155, which inhibits expression of HIF, which insults cancerous cells and induces apoptosis. In summary, MeJA indicates unique and imperative aspects concerning the assimilated biological roles against oxidative stress, viz. reducing infiltration of inflammatory cells and their activation, inhibition of proinflammatory mediators (IL, TNF-), LOX and COX,

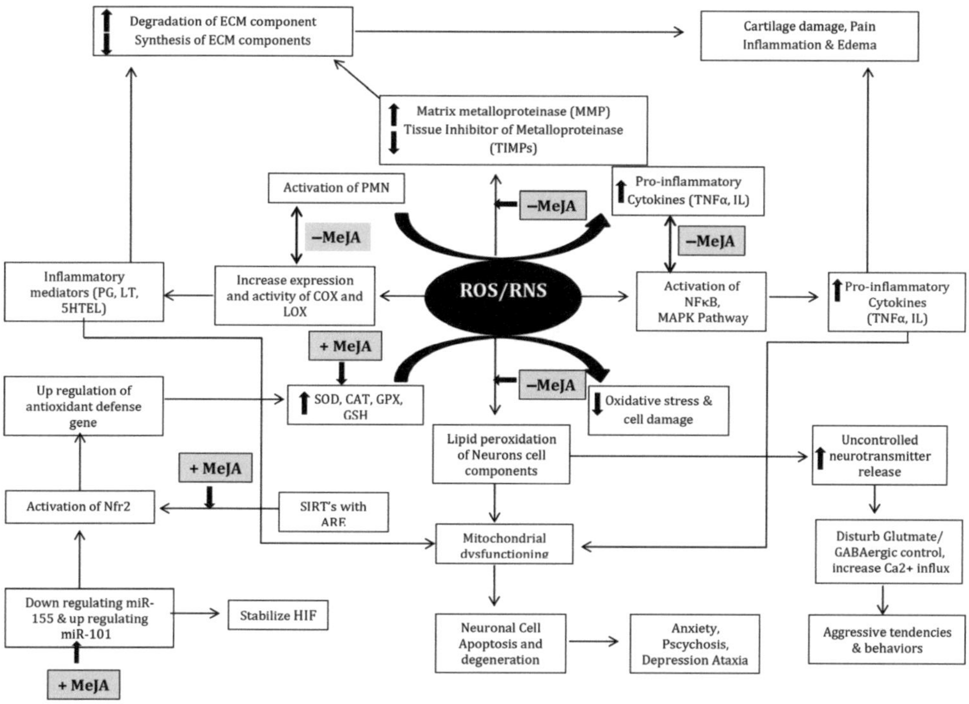


**Figure 1.**
*MeJA multifunctional role in oxidative stress and molecular interactions in antioxidant defense mechanism.*

downregulation of NF-B and MAPK transcription pathways, downregulation of miRNA-155, and upregulation of miRNA-101 and Nfr2 pathway (**Figure 1**).

Reactive oxygen species (ROS)/reactive oxygen species (RNS) initiates signaling inflammatory pathways NF-κB (nuclear factor-kB)/MAPK (Mitogen-activated protein kinase) and induces synthesis of cytokines (IL and TNF-α), (MAPK) responsible to enhance activity of COX and increases the synthesis of pro-inflammatory mediators. Activity of COX (cyclooxygenase)/LOX (Lipoxygenase) is enhanced by ROS/RNS and induces synthesis of inflammatory mediators like (PGs, LTs, 5HTEL). The formed inflammatory mediators contribute to destructive effects on the different cells. Lipid peroxidation caused by ROS/RNS leads to damage to neuronal cells component and disturbs mitochondria functioning and induces neuronal cell degeneration and death and develops anxiety, psychosis, depression etc. The disturbed neuronal cell functioning cause's uncontrolled release of neurotransmitter and causes imbalance between inhibitory (GABAergic) and excitatory (Glutmate) neuronal mechanism leads to aggressive tendencies & behaviors. On induction of inflammation PMN (polymorphonuclear neutrophil) activates and are important source for generation of ROS/RNS which further participate in inflammatory cascades. ROS/RNS increases activity of destructive MMP (matrix metalloproteinase) whereas protective tissue TIMPs (Inhibitor of metalloproteinase) is diminished. MMP selectively degrades the component of cartilage ECM (Extra cellular matrix) and causes cartilage damage, pain inflammation due to wear and tear of joints. MeJAregulates miR-155, miR-101 as well as sirtuins (SIRTs) antioxidant and redox signaling pathway leads to the upregulation of the Nfr2 (nuclear factor erythroid 2-related 2) activity. Up regulation of the Nfr2 increases gene transcription for antioxidant enzymes and reduces cellular oxidative stress.

## 4. Conclusion

PTH jasmonic acid and its derivatives like MeJA are important in the survival of plants in biotic and abiotic stressful conditions as well as have proved their effectiveness in the treatment of several clinical ailments. An important consideration has been pointed out to oxidative stress-mediated by ROS in the development of several pathological conditions like cardiovascular, metabolic, psychosis, and neurodegenerative disorders, cancer, etc. MeJA is not only effective in plants to relieve oxidative stress, but also effectively relieves the same in human beings. ROS activates several pathways like the NF-B and MAPK signaling pathway, increases the activity of inflammatory (PG, LT, 5HTEL) and pro-inflammatory mediators (IL, TNF-), triggers classes of degradative enzymes, disturbs cellular redox status and depletes antioxidant enzymes, induces lipid peroxidation of important cell components, and disturbs cellular normal physiology and construction. Targeting ROS/RNS by antioxidant molecules or disabling signaling pathways activated by ROS are important concerns for treatment options in severe diseases and disorders. MeJA has shown a prominent role in controlling and neutralizing signaling pathways like NF-κB and MAPK also effectively reduces the activity of inflammatory mediators, oxidative stress, and protects the cell and its components from ROS. On the other hand, methyl jasmonate has a positive interaction with miRNA-101 which activates Nfr2mediated upregulation of antioxidant defense genes for SOD, CAT, and GSH enzymes as well as indirectly boosts SIRT antioxidant and redox signaling pathways. Considering the potent role of MeJA and significant interference in oxidative stress and facilitated disease causative pathways, it could be an influential candidate in the treatment of numerous pathological conditions. Several molecules have been screened from natural and synthetic sources for their potential antioxidant benefits. Among them, MeJA has a multifaceted role against oxidative stress-mediated cellular damage. In conclusion, phytohormones like MeJA may be a protruding candidate for new drug discovery and a highly promising molecule for the pharmacotherapy of severe diseases or disorders.

## Acknowledgements

We would like to acknowledge the Management and Principle of ASPM's K. T. Patil College of, Osmanabad and JSPM's Charak College of Pharmacy and Research, Pune for providing necessary facilities and infrastructure. The current chapter is not funded by any Government/Non-government organizations.

## Conflict of interest

The authors declare no conflict of interest.

## Acronyms and abbreviations

| | |
|---|---|
| JA | Jasmonic Acid |
| CK | Cytokinins |
| GA | Gibberellins |
| ABA | Abscisic Acid |
| ET | Ethylene |
| SA | Salicylic Acid |

| | |
|---|---|
| PTH | Phytohormones |
| MeJA | Methyl Jasmonate |
| NF-B | Nuclear Factor Kappa-Light-chain-enhancer of Activated B cells |
| MAPK | Mitogen-Activated Protein Kinase |
| IL | INTERLEUKIN |
| TNF-α | Tumor Necrosis Factor alpha |
| COX | Cyclo-Oxygenase |
| LOX | LIPO-OXYGENASE |
| miRNA | microrna |
| Nfr2 | Nuclear Factor Erythroid |
| SIRTs | Sirtuins |
| ROS | Reactive Oxygen Species |
| RNS | Reactive Nitrogen Species |
| $H_2O_2$ | Hydrogen Peroxide |
| ROOR | Organic Hydroperoxide |
| NO | Nitric Oxide |
| O- | Superoxide |
| •OH | Hydroxyl Radicals |
| SOD | Superoxide Dismutase |
| CAT | Catalase |
| GSH | Glutathione |
| GR | Glutathione Reductase |
| MDA | Malondialdehyde |
| IkB | Inhibitory Protein B |
| PG | Prostaglandins |
| LT | Leukotrienes |
| 5-HETE | 5-Hydroxyeicosatetraenoic Acid |
| AD | Alzheimer's Disease |
| Ach | Acetylcholine |
| AChE | Acetylcholine Esterase |
| NSAIDs | Non-Steroidal Anti-inflammatory drugs |
| LPS | Lipopolysaccharides |
| PLA2 | Phospholipase A2 |
| AA | Arachidonic Acid |
| DHA | Docosahexaenoic Acid |
| GABA | Gamma-Aminobutyric Acid |
| NOS | Nitric Oxide Synthase |
| ECM | Extracellular Matrix |
| TIMPs | Tissue Inhibitors of Metalloproteinases |
| MMP | Matrix Metalloproteinases |
| HOCl | Hypochlorous Acid |
| EZH2 | Enhancer of Zeste Homolog 2 |
| ARE | Antioxidant Response Elements |
| HIFs | Hypoxia Induced Factors |

**Author details**

Gunjegaonkar Shivshankar M.[1*], Joshi Amol A.[1], Wankhede Sagar B.[2], Siraskar Balasaheb D.[3], Merekar Abhijit N.[4] and Shinde Sachin D.[5]

1 Department of Pharmacology, ASPM's K. T. Patil College of Pharmacy, Osmanabad, Maharashtra, India

2 Department of Pharmaceutical Chemistry, JSPM's Charak College of Pharmacy and Research, Wagholi, Pune, Maharashtra, India

3 Department of Pharmacology, Shri Vivekananda Nursing Home Trusts College of B. Pharmacy, Ahmednagar, Maharashtra, India

4 Dr. Vithalrao Vikhe Patil Foundation College of Pharmacy, Ahmednagar, Maharashtra, India

5 Department of Pharmacology, Shri. R. D. Bhakt College of Pharmacy, Jalna, Maharashtra, India

*Address all correspondence to: gunjeshiv@gmail.com

## References

[1] Apel K, Hirt H. Reactive oxygen species: Metabolism, oxidative stress, and signal transduction. Annual Review of Plant Biology. 2004;**55**:373-399

[2] Finkel T, Holbrook NJ. Oxidants, oxidative stress and the biology of aging. Nature. 2000;**408**(6809):239

[3] Salvemini D, Cuzzocrea S. Oxidative stress in septic shock and disseminated intravascular coagulation. Free Radical Biology and Medicine. 2002;**33**(9): 1173-1185

[4] Farber JL. Mechanisms of cell injury by activated oxygen species. Environmental Health Perspectives. 1994;**102**(10):17-24

[5] Mattill H. Antioxidants. Annual Review of Biochemistry. 1974;**16**(1): 177-192

[6] Brüne B, Zhou J, Von Knethen A. Nitric oxide, oxidative stress, and apoptosis. Kidney International. 2003; **63**:S22-S24

[7] Noori S. An overview of oxidative stress and antioxidant defensive system. Open Access Scientific Reports. 2012;**1**(8):1-9

[8] Rahman T, et al. Oxidative Stress and Human Health. Advances in Bioscience and Biotechnology. 2012;**3**:1-24

[9] Sevanian A, Hochstein P. Mechanisms and consequences of lipid peroxidation in biological systems. Annual Review of Nutrition. 1985;**5**(1):365-390

[10] Evans JL et al. Are oxidative stress—Activated signaling pathways mediators of insulin resistance and β-cell dysfunction? Diabetes. 2003;**52**(1):1-8

[11] West IC. Radicals and oxidative stress in diabetes. Diabetic Medicine. 2000;**17**(3):171-180

[12] Lefer DJ, Granger DN. Oxidative stress and cardiac disease. American Journal of Medicine. 2000;**109**(4): 315-323

[13] Singh U, Devaraj S, Jialal I. Vitamin E, oxidative stress, and inflammation. Annual Review of Nutrition. 2005;**25**:151-174

[14] Ding Q, Dimayuga E, Keller JN. Proteasome regulation of oxidative stress in aging and age-related diseases of the CNS. Antioxidants & Redox Signaling. 2006;**8**(1-2):163-172

[15] Hiran TS, Moulton PJ, Hancock JT. Detection of superoxide and NADPH oxidase in porcine articular chondrocytes. Free Radical Biology and Medicine. 1997;**23**(5):736-743

[16] Fubini B, Hubbard A. Reactive oxygen species (ROS) and reactive nitrogen species (RNS) generation by silica in inflammation and fibrosis. Free Radical Biology and Medicine. 2003; **34**(12):1507-1516

[17] John Aitken R, Clarkson JS, Fishel S. Generation of reactive oxygen species, lipid peroxidation, and human sperm function. Biology of Reproduction. 1989;**41**(1):183-197

[18] Matés JM, Pérez-Gómez C, De Castro IN. Antioxidant enzymes and human diseases. Clinical Biochemistry. 1999;**32**(8):595-603

[19] Oberley LW, Buettner GR. Role of superoxide dismutase in cancer: A review. Cancer Research. 1979;**39**(4): 1141-1149

[20] Vaziri ND et al. Oxidative stress and dysregulation of superoxide dismutase and NADPH oxidase in renal insufficiency. Kidney International. 2003;**63**(1):179-185

[21] Chakravarthi S, Jessop CE, Bulleid NJ. The role of glutathione in disulphide bond formation and endoplasmic-reticulum-generated oxidative stress. EMBO Reports. 2006;**7**(3):271-275

[22] Lieber CS. Role of oxidative stress and antioxidant therapy in alcoholic and nonalcoholic liver diseases. In: Advances in Pharmacology. Amsterdam, Netherlands: Elsevier; 1996. pp. 601-628

[23] Gogos CA et al. Pro-versus anti-inflammatory cytokine profile in patients with severe sepsis: A marker for prognosis and future therapeutic options. The Journal of Infectious Diseases. 2000;**181**(1):176-180

[24] Li Q, Verma IM. NF-κB regulation in the immune system. Nature Reviews Immunology. 2002;**2**(10):725-734

[25] Adcock IM et al. Oxidative stress induces NFκB DNA binding and inducible NOS mRNA in human epithelial cells. Biochemical and Biophysical Research Communications. 1994;**199**(3):1518-1524

[26] Ghosh J et al. Taurine prevents arsenic-induced cardiac oxidative stress and apoptotic damage: Role of NF-κB, p38 and JNK MAPK pathway. Toxicology and Applied Pharmacology. 2009;**240**(1):73-87

[27] Johnstone M, Gearing AJ, Miller KM. A central role for astrocytes in the inflammatory response to β-amyloid; chemokines, cytokines and reactive oxygen species are produced. Journal of Neuroimmunology. 1999;**93**(1-2):182-193

[28] Gunjegaonkar S, Shanmugarajan T. Methyl jasmonate a stress phytohormone attenuates LPS induced in vivo and in vitro arthritis. Molecular Biology Reports. 2019;**46**(1):647-656

[29] Vaquero E et al. Localized pancreatic NF-κB activation and inflammatory response in taurocholate-induced pancreatitis. American Journal of Physiology-Gastrointestinal and Liver Physiology. 2001;**280**(6):G1197-G1208

[30] Song P, Zhao Z-Q, Liu X-Y. Expression of IL-2 receptor in dorsal root ganglion neurons and peripheral antinociception. Neuroreport. 2000;**11**(7):1433-1436

[31] Shanmugarajan T. Potential of plant stress hormone methyl Jasmonate against lipopolysaccharide attenuated oxidative stress and arthritis in experimental animals. International Journal of Green Pharmacy (IJGP). 2018;**12**(3):561-572

[32] Umukoro S et al. Anti-inflammatory and membrane stabilizing properties of methyl jasmonate in rats. Chinese Journal of Natural Medicines. 2017;**15**(3):202-209

[33] Umukoro S, Olugbemide AS. Antinociceptive effects of methyl jasmonate in experimental animals. Journal of Natural Medicines. 2011;**65**(3-4):466-470

[34] Kim M et al. Methyl jasmonate inhibits lipopolysaccharide-induced inflammatory cytokine production via mitogen-activated protein kinase and nuclear factor-κB pathways in RAW 264.7 cells. Die Pharmazie-An International Journal of Pharmaceutical Sciences. 2016;**71**(9):540-543

[35] Koedam EL et al. Early-versus late-onset Alzheimer's disease: More than age alone. Journal of Alzheimer's Disease. 2010;**19**(4):1401-1408

[36] Schofield PW et al. The age at onset of Alzheimer's disease and an intracranial area measurement: A relationship. Archives of Neurology. 1995;**52**(1):95-98

[37] Spillantini MG, Goedert M. Tau protein pathology in neurodegenerative

diseases. Trends in Neurosciences. 1998;**21**(10):428-433

[38] Nunomura A et al. Involvement of oxidative stress in Alzheimer disease. Journal of Neuropathology & Experimental Neurology. 2006;**65**(7): 631-641

[39] Zhu X et al. Causes of oxidative stress in Alzheimer disease. Cellular and Molecular Life Sciences. 2007;**64**(17):2202-2210

[40] Markesbery W, Lovell M. Four-hydroxynonenal, a product of lipid peroxidation, is increased in the brain in Alzheimer's disease. Neurobiology of Aging. 1998;**19**(1):33-36

[41] Williams TI et al. Increased levels of 4-hydroxynonenal and acrolein, neurotoxic markers of lipid peroxidation, in the brain in mild cognitive impairment and early Alzheimer's disease. Neurobiology of Aging. 2006;**27**(8):1094-1099

[42] Yamada K, Nabeshima T. Animal models of Alzheimer's disease and evaluation of anti-dementia drugs. Pharmacology & Therapeutics. 2000;**88**(2):93-113

[43] Zarifkar A et al. *Agmatine* prevents LPS-induced spatial memory impairment and hippocampal apoptosis. European Journal of Pharmacology. 2010;**634**(1-3):84-88

[44] Kitazawa M et al. Lipopolysaccharide-induced inflammation exacerbates tau pathology by a cyclin-dependent kinase 5-mediated pathway in a transgenic model of Alzheimer's disease. Journal of Neuroscience. 2005;**25**(39):8843-8853

[45] Umukoro S, Eduviere A. Methyl jasmonate attenuated lipopolysaccharide-induced memory dysfunction through inhibition of neuroinflammatory markers and beta-amyloid generation in mice. European Neuropsychopharmacology. 2016;**2**(26):S647-S648

[46] Eduviere AT et al. Methyl jasmonate enhances memory performance through inhibition of oxidative stress and acetylcholinesterase activity in mice. Life Sciences. 2015;**132**:20-26

[47] Eduviere AT, Omorogbe O, Umukoro S. Methyl jasmonate ameliorates memory deficits in mice exposed to passive avoidance paradigm. Journal of Neuroscience Research. 2017;**1**:6

[48] Annafi OS, Umukoro S, Eduviere AT. Evaluation of the anticonvulsant and anxiolytic potentials of methyl jasmonate in mice. Scientia Pharmaceutica. 2014;**82**(3):643-654

[49] Chrousos GP, Gold PW. The concepts of stress and stress system disorders: Overview of physical and behavioral homeostasis. JAMA. 1992;**267**(9):1244-1252

[50] Bouayed J, Rammal H, Soulimani R. Oxidative stress and anxiety: Relationship and cellular pathways. Oxidative Medicine and Cellular Longevity. 2009;**2**(2):63-67

[51] Wilensky U. What is normal anyway? Therapy for epistemological anxiety. Educational Studies in Mathematics. 1997;**33**(2):171-202

[52] Halliwell B. Oxidative stress and neurodegeneration: Where are we now? Journal of Neurochemistry. 2006; **97**(6):1634-1658

[53] Bouayed J et al. Positive correlation between peripheral blood granulocyte oxidative status and level of anxiety in mice. European Journal of Pharmacology. 2007;**564**(1-3):146-149

[54] Hakonen AH et al. Mitochondrial DNA polymerase W748S mutation:

A common cause of autosomal recessive ataxia with ancient European origin. The American Journal of Human Genetics. 2005;**77**(3):430-441

[55] Suomalainen A et al. Multiple deletions of mitochondrial DNA in several tissues of a patient with severe retarded depression and familial progressive external ophthalmoplegia. The Journal of Clinical Investigation. 1992;**90**(1):61-66

[56] Crews FT et al. Alcohol-induced neurodegeneration: When, where and why? Alcoholism: Clinical and Experimental Research. 2004;**28**(2): 350-364

[57] Sun GY et al. Phospholipase A2 in the central nervous system implications for neurodegenerative diseases. Journal of Lipid Research. 2004;**45**(2): 205-213

[58] Caro AA, Cederbaum AI. Role of cytochrome P450 in phospholipase A2-and arachidonic acid-mediated cytotoxicity. Free Radical Biology and Medicine. 2006;**40**(3):364-375

[59] Suganuma H et al. Maternal docosahexaenoic acid-enriched diet prevents neonatal brain injury. Neuropathology. 2010;**30**(6):597-605

[60] Mattson MP, Camandola S. NF-κB in neuronal plasticity and neurodegenerative disorders. The Journal of Clinical Investigation. 2001;**107**(3):247-254

[61] Terai K et al. Enhancement of immunoreactivity for NF-κB in human cerebral infarctions. Brain Research. 1996;**739**(1-2):343-349

[62] Annafi OS et al. Probable mechanisms involved in the antipsychotic-like activity of methyl jasmonate in mice. Naunyn-Schmiedeberg's Archives of Pharmacology. 2017;**390**(9):883-892

[63] Zalcman SS, Siegel A. The neurobiology of aggression and rage: Role of cytokines. Brain, Behavior, and Immunity. 2006;**20**(6):507-514

[64] Inagaki TK et al. Inflammation selectively enhances amygdala activity to socially threatening images. NeuroImage. 2012;**59**(4):3222-3226

[65] Lebel CP, Bondy SC. Oxygen radicals: Common mediators of neurotoxicity. Neurotoxicology and Teratology. 1991;**13**(3):341-346

[66] Gidron Y et al. The relation between psychological factors and DNA-damage: A critical review. Biological Psychology. 2006;**72**(3):291-304

[67] Dawson TM, Dawson VL, Snyder SH. Molecular mechanisms of nitric oxide actions in the brain a. Annals of the New York Academy of Sciences. 1994;**738**(1):76-85

[68] Hovatta I, Juhila J, Donner J. Oxidative stress in anxiety and comorbid disorders. Neuroscience Research. 2010;**68**(4):261-275

[69] Umukoro S, Eduviere AT, Aladeokin AC. Anti-aggressive activity of methyl jasmonate and the probable mechanism of its action in mice. Pharmacology Biochemistry and Behavior. 2012;**101**(2):271-277

[70] Díaz-Prado S et al. Potential use of the human amniotic membrane as a scaffold in human articular cartilage repair. Cell and Tissue Banking. 2010;**11**(2):183-195

[71] Panseri S et al. Osteochondral tissue engineering approaches for articular cartilage and subchondral bone regeneration. Knee Surgery, Sports Traumatology, Arthroscopy. 2012;**20**(6):1182-1191

[72] Gibson J et al. Oxygen and reactive oxygen species in articular cartilage:

Modulators of ionic homeostasis. Pflügers Archiv-European Journal of Physiology. 2008;**455**(4):563-573

[73] Martin JA et al. Effects of oxidative damage and telomerase activity on human articular cartilage chondrocyte senescence. The Journals of Gerontology Series A: Biological Sciences and Medical Sciences. 2004;**59**(4):B324-B336

[74] Billinghurst RC et al. Enhanced cleavage of type II collagen by collagenases in osteoarthritic articular cartilage. The Journal of Clinical Investigation. 1997;**99**(7):1534-1545

[75] Khan IM et al. Oxidative stress induces expression of osteoarthritis markers procollagen IIA and 3B3 (−) in adult bovine articular cartilage. Osteoarthritis and Cartilage. 2008;**16**(6):698-707

[76] Murphy G. Tissue inhibitors of metalloproteinases. Genome Biology. 2011;**12**(11):233

[77] Lotz M. The role of nitric oxide in articular cartilage damage. Rheumatic Disease Clinics of North America. 1999;**25**(2):269-282

[78] Goldring MB. Osteoarthritis and cartilage: The role of cytokines. Current Rheumatology Reports. 2000;**2**(6): 459-465

[79] Schneider N et al. Synoviocytes, not chondrocytes, release free radicals after cycles of anoxia/re-oxygenation. Biochemical and Biophysical Research Communications. 2005;**334**(2):669-673

[80] Jovanovic DV et al. Nitric oxide induced cell death in human osteoarthritic synoviocytes is mediated by tyrosine kinase activation and hydrogen peroxide and/or superoxide formation. The Journal of Rheumatology. 2002;**29**(10):2165-2175

[81] Zhang L et al. MicroRNA-217 is involved in the progression of atherosclerosis through regulating inflammatory responses by targeting sirtuin 1. Molecular Medicine Reports. 2019;**20**(4):3182-3190

[82] Engedal N et al. From oxidative stress damage to pathways, networks, and autophagy via microRNAs. Oxidative Medicine and Cellular Longevity. 2018;**2018**:1-16

[83] Peng Z, Zhang Y. Methyl jasmonate induces the apoptosis of human colorectal cancer cells via downregulation of EZH2 expression by microRNA-101. Molecular Medicine Reports. 2017;**15**(2):957-962

[84] Wang Y et al. Methyl jasmonate sensitizes human bladder cancer cells to gambogic acid-induced apoptosis through down-regulation of EZH 2 expression by miR-101. British Journal of Pharmacology. 2014;**171**(3):618-635

[85] Cesari IM et al. Methyl jasmonate: Putative mechanisms of action on cancer cells cycle, metabolism, and apoptosis. International Journal of Cell Biology. 2014;**2014**:1-12

[86] Kastl L et al. TNF-α mediates mitochondrial uncoupling and enhances ROS-dependent cell migration via NF-κB activation in liver cells. FEBS Letters. 2014;**588**(1):175-183

[87] Dinkova-Kostova AT, Abramov AY. The emerging role of Nrf2 in mitochondrial function. Free Radical Biology and Medicine. 2015;**88**:179-188

[88] Guo Y et al. Epigenetic regulation of Keap1-Nrf2 signaling. Free Radical Biology and Medicine. 2015;**88**:337-349

[89] Yu D-S et al. Salvianolic acid A ameliorates the integrity of blood-spinal cord barrier via miR-101/Cul3/Nrf2/HO-1 signaling pathway. Brain Research. 2017;**1657**:279-287

[90] Sá-Nakanishi AB et al. Anti-inflammatory and antioxidant actions

of methyl jasmonate are associated with metabolic modifications in the liver of arthritic rats. Oxidative Medicine and Cellular Longevity. 2018;**2018**:1-15

[91] Yamakuchi M. MicroRNA regulation of SIRT1. Frontiers in Physiology. 2012;**3**:68

[92] Choi S-E, Kemper JK. Regulation of SIRT1 by microRNAs. Molecules and Cells. 2013;**36**(5):385-392

[93] Nguyen T, Sherratt PJ, Pickett CB. Regulatory mechanisms controlling gene expression mediated by the antioxidant response element. Annual Review of Pharmacology and Toxicology. 2003;**43**(1):233-260

[94] Haigis MC, Sinclair DA. Mammalian sirtuins: Biological insights and disease relevance. Annual Review of Pathology: Mechanisms of Disease. 2010;**5**:253-295

[95] Merksamer PI et al. The sirtuins, oxidative stress and aging: An emerging link. Aging (Albany NY). 2013;**5**(3): 144

[96] Gunjegaonkar S, Shanmugarajan T. Molecular mechanism of plant stress hormone methyl jasmonate for its anti-inflammatory activity. Plant Signaling & Behavior. 2019;**14**(10): e1642038

[97] Umukoro S et al. Evaluation of adaptogenic-like property of methyl jasmonate in mice exposed to unpredictable chronic mild stress. Brain Research Bulletin. 2016;**121**:105-114

[98] Pialoux V et al. Relationship between oxidative stress and HIF-1α mRNA during sustained hypoxia in humans. Free Radical Biology and Medicine. 2009;**46**(2):321-326

[99] Nardinocchi L et al. Zinc downregulates HIF-1α and inhibits its activity in tumor cells in vitro and in vivo. PLoS One. 2010:**5**(12):1-8

[100] Serocki M et al. miRNAs regulate the HIF switch during hypoxia: A novel therapeutic target. Angiogenesis. 2018;**21**(2):183-202

## Section 2

# Phytohormones Applications

Chapter 5

# Key Aspects of Plant Hormones in Agricultural Sustainability under Climate Change

*Muhammad Amjad Bashir, Qurat-Ul-Ain Raza, Hafiz Muhammad Ali Raza, Muhammad Umair Sial, Abdur Rehim, Kashif Ali Khan, Muhammad Ijaz and Muhammad Wasif*

## Abstract

Climate change is an emerging issue for modern agriculture and has generated biotic and abiotic stresses for plants such as cold, high temperature, heat, drought, uneven rainfall, and UV radiations. In addition to these, serious stress factors are emerging related to water availability, nutrient cycling, salinity-sodicity, and pest attacks. In recent years, such phenomena have attracted the research community to avoid the fatal influence of climate change on crop production and obtain more food helping in fulfillment of increasing food demand of population surge. The anthropogenic activities in the agroecosystem are among the major causes for global warming and proportion in climate change. Therefore, it is assumed that identifying various plant hormones and their utilization to improve plant metabolic activities would help maintain plant growth, survival, and production under severe climate change circumstances. This chapter focuses on identifying the key aspects of plant hormones to retard the negative impacts of climate change and support sustainable agriculture.

**Keywords:** plant stress, abiotic factors, biostimulants, plant growth regulators, agricultural production

## 1. Introduction

Issues generated from changing climate are trending throughout the globe and have altered various Earth ecosystem processes [1]. Climatic variability assumes about 60% of yield variability, and therefore, it is considered a critical factor to influence crop productivity and farmer's income. In addition to crop yield and productivity, different natural resources such as land and water are also exploited and have a crucial impact on agricultural production [2]. The primary reason for the substantial changes in climate in the last few decades is the excessive anthropogenic activities that transformed the composition of the global atmosphere. Since 1750, the greenhouses gases (GHGs) concentration increased, including methane (150%), carbon dioxide (40%), and nitrous oxide (20%), respectively [3].

The growing population has increased the demand for food that has resulted in intensive agricultural practices such as excessive fertilizer usage, manipulation of water resources, and livestock generation. Furthermore, such agricultural activities result in producing GHGs, thus polluting natural resources. Additionally, climate change adversely degrades the land resulting in increased desertification and having nutrient-deficient soils. The threat of growing land degradation day by day is another issue observed worldwide. Global Assessment of Land Degradation and Improvement (GLADA) reported that a quarter of the global land has not to be categorized as degraded. Climate change and anthropogenic activities are the key factors to deteriorate the 15 billion tons of fertile soil every year, and its 1.5 billion people are also affected [1].

The agriculture sector is also prone to climate change as it is sensitive to weather conditions and causes massive impacts on economics. The crop yield is also affected by changes in climatic events, including temperature and rainfall. Increase in temperature, changes in precipitation patterns, and $CO_2$ fertilization vary due to the crop, location, and magnitude of change in the parameters (**Figure 1**). The increase in precipitation is likely to reduce the temperature, which ultimately reduces the crop yield. Moreover, humidity and wind speed also impact crop productivity and insect pests and diseases, which are more active in humid and warmer conditions [3]. Climate change has also contributed to drought, increasing sea levels, intense storms, and floods, resulting in land degradation [4].

A collection of small molecules having a different structure that contributes to improving various plant growth and development processes and showing significant results against biotic and abiotic stresses are termed as plant hormones. Moreover, plant hormones regulate different mechanisms such as seed germination, cell differentiation, cell proliferation, senescence, stem elongation, organ formation and response to drought, pest attack, and wounding. Thus, plant growth and productivity, size, and architecture are controlled by plant hormones. In addition, it can be a possible solution to improve agricultural productivity and solve the worldwide food storage issues [5, 6].

Plant hormones can also be defined as the plant growth regulators either produced inside or by the plants, whereas plant growth regulators are synthetic

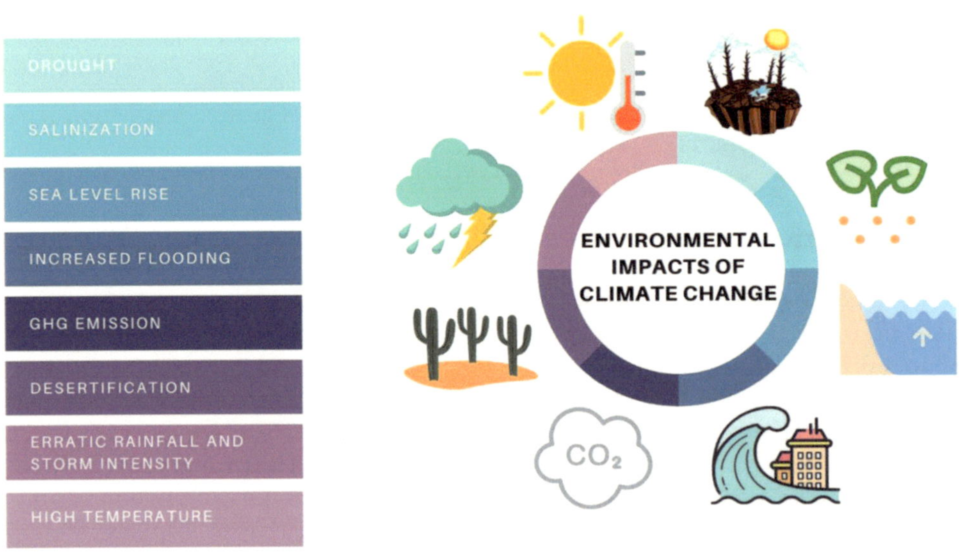


**Figure 1.**
*Environmental impacts resulting from climate change.*

materials that can alter plant biological processes and improve its growth [6, 7]. Major classical hormones included in the plant growth regulator category are abscisic acid, cytokinin, ethylene, auxin, jasmonate, gibberellin, salicylic acid, brassinosteroids—however, many more there to be discovered. For the last few years, the understanding of the plant hormones and their association with various plant processes, the mechanism behind them, and their signaling roles has been progressed [8].

In this chapter, we have aimed to understand plant hormones as a tool to achieve agricultural sustainability to tackle climate change. The chapter consists of eight sections, which describes the introduction (Section 1), interaction of plant hormones and plant physiology (Section 2), role of plant hormones in adaptation to salt stress (Section 3), participation of plant hormones in heat and cold stress tolerance (Section 4), the response of plant hormones to drought stress (Section 5), the involvement of phytohormones in resetting plant-pest interaction (Section 6), future prospects and challenges (Section 7), and the conclusion (Section 8), respectively.

## 2. Interaction of plant hormones with plant physiology

Plant hormones are chemicals and should not be considered nutrients. These hormones are programmed to respond at a specific time or growth stage, or cycle. However, before or after that specific stage, the impact of hormones will diminish [9]. Due to climate change, plants face various biotic and abiotic stresses, including salinity, cold and heat stress, drought, insects, and pathogens attack that adversely impact plant growth and yield. However, plant hormones in a small amount not only protect the plant from environmental stresses but also improve crop yield [10], as described in **Table 1**.

Plant hormones help to attain resistance in plants against induced stress (**Figure 2**). Abscisic acid is related to the plant stress responses such as cold, drought, salinity, and plant growth, including seed dormancy, embryo maturation, flower induction, and

| Plant hormone | Identified primary function | References |
|---|---|---|
| Abscisic acid | Impacts various cellular processes (i.e.,) seed maturation and germination, stomatal movement, and leaf senescence. | [11] |
| Cytokinin | Influence cell division in roots and shoots, delay tissues senescence. | [12] |
| Ethylene | Promote, inhibit, or induce growth and development of leaves, flowers, and fruits. | [13] |
| Auxins | It affects plant development as cell division, elongation, differentiation, flowering, and senescence. | [14] |
| Jasmonates | It contributes to developing plant roots and reproductive systems, develops defense mechanisms, and triggers gene expression. | [10] |
| Gibberellin | Facilitates plant growth, cell elongation, and development of flowers, fruits, and seeds. | [15] |
| Salicylic acid | Develops resistance in plants and stimulates the production of antioxidants. | [16] |
| Brassinosteroids | Regulate various physiological processes such as cell expansion and proliferation, vascular differentiation, timing senescence, and male fertility. | [17] |

**Table 1.**
*Pivotal roles of plant hormones in plant growth and development.*

**Figure 2.**
*Plant hormones enable the plants to sustain against environmental stresses.*

pest attack [18]. Cytokinin is a hormone produced by plants as well as nearby insects and pathogens. However, it induces defense mechanisms in plants upon pathogen attack and impacts the plant physiological traits, including apical dominance, seed germination, leaf senescence, flower, and fruit development, whereas type of molecules of cytokinin hormones varies depending upon the plant growth stages and environmental conditions [19].

Ethylene has a significant role in leaf growth, development, senescence. However, the response depends on the concentration of hormones and the plant species. Therefore, the effect of ethylene on plant physiology might be an independent response or dependent on the interaction with other hormones. Moreover, ethylene has been shown to play an antagonistic role to auxins in the abscission of various organs. Furthermore, ethylene is vital for fruit ripening, plant aging, and protecting the seed until maturity [13]. On the other hand, auxin is a growth hormone and involves in plant growth and development through cell elongation and expansion. It also regulates organ development, shoot, root branching, gravitropism, and phototropism [20].

Jasmonic acid is a plant hormone and is involved in plant defense mechanisms. Thus, the plant can perform effectively under stressful environmental conditions. Its interaction with ethylene improves root development and anthocyanin accumulation that could be related to its tolerance against abiotic stresses [10]. Furthermore, it is involved in the inhibition of seed germination. The mechanism behind the role of Jasmonic acid varies depending upon the environmental constraints [21]. Gibberellin is required by the plants at multiple stages and contributes to germination, the transition to flowering, and flower development, root elongation, and fruit development [22].

Salicylic acid is also a defense-boosting hormone as its level increases with pest attack [21]. In addition, brassinosteroids regulate specific gene sets to help plants sustain during unfavorable environmental conditions [23]. Moreover, it is also involved in various metabolic processes, including osmotic regulation, plant-water interaction, photosynthesis, nitrogen metabolism, and antioxidant metabolism [24].

## 3. Role of plant hormone in adaptation to salt stress

With the increasing population, it is challenging to fulfill the food requirements, as climate change has threatened the sustainability and productivity of the agriculture production system [25]. Different types of abiotic stresses are present in nature, affecting the various processes in plants and indicating the unique and complex response against the stresses, depending on multiple factors such as degree of plasticity, including many morpho-physiological, cellular, and anatomical [26]. According to an estimate, about 6% of the world's total land area is affected by salinity. Among various biotic and abiotic stresses, salinity also plays a vital role in decreasing crop growth and productivity [27] that has been intensified due to poor irrigation practices, increasing population, and industrial pollution [28]. Some other types of stress, including osmotic stress, ionic stress, and oxidative stress, occur in the plant because of salt stress. Consequently, to survive under such conditions, plants need to rely on such critical pathways, which help the plants reestablish the various processes, including ionic, osmotic, and reactive oxygen species [27]. It is necessary for the plants that grow under salt stress conditions to re-adjust the different biochemical and physiological processes involved under salt stress, and plants have to adjust their physiological and biochemical processes, engaged in modifying not only the ionic but also the osmotic homeostasis [29].

Several studies have shown that every hormone present in the plant does not play only a biological role in the plant, but also plays a vital role in various important stages, such as tissue formation and other physiological processes [30, 31]. These phytohormones can act either near to or remove from the site of synthesis to regulate responses to environmental stimuli or genetically programmed developmental changes [32]. So, hormones play a vital role in facilitating plant response to various abiotic stresses. The plant may try to spurt or survive under stress conditions and may also decrease plant growth. Thus, the plant focuses on its resources on withstanding the stress [33]. Abiotic stresses may cause a different type of damage, often leading to adjustments in production, supply, and signal transductions of growth along with stress hormones that can promote definite protective mechanisms [26]. Awareness of a stress signal generates the signal transduction pours in plants with hormones acting as the baseline transducers [34].

## 4. Participation of plant hormones in heat and cold stress tolerance

Plants have established a different but extraordinary ability to acclimatize the harsh environment and flourish in habitats categorized by abiotic stresses, for example, temperature extremes [35]. Both exogenous and endogenous environmental factors are involved in regulating plant growth and development [8]. All the exogenous environmental factors (temperature, light, moisture, and atmospheric CO2) must be present at an optimal level to regulate the metabolic process. Temperature is one of the most important factors among all environmental factors because it plays a dynamic role in regulating the phenological development of a crop plant [36]. Temperature beyond the "physiological optimum" that affects a plant's optimal growth is usually known as "high temperature" for that particular plant. The participation of hormones in the plant response to heat stress can be examined in many different ways. Associate approaches such as short-term heat shock with enormously high-temperature and heat acclimation study reveal plants to minor heat stress previously imposing great heat stress and long-term high-temperature treatment [37].

Some important crops (annual) such as wheat, oats, barley, and pea show a substantial degree of intrinsic freezing tolerance, which can be further enhanced by using complex signaling events. On the other hand, some species such as maize, rice, and tomato, which belong to temperate or tropical zones, may face severe damage at a chilling temperature [35]. Hormones are chemical messengers and low-molecular-weight complexes that transfer essential information from a production site to the place of action. Regulation of hormonal homeostasis is taken up by different processes such as biosynthesis, catabolism, and their transformation from one place to another. At the same time, the sensitivity is produced due to the presence and response of dedicated receptors determined by the receiving tissue, which pledge the signal transduction action to change the cellular process [38].

Various types and classes of hormones are available that are involved in the different and overlapping functions. Furthermore, hormones manifest different synergistic and antagonistic effects on the synthesis and signaling productions of the other hormones, generating a composite network of hormonal relations [38]. The hormonal signaling system assimilates the peripheral information into endogenous development programs and initiates the stress-responsive pathway leading toward resistance. Hence, it may not be surprising that plants utilize phytohormones to signal cold stress. So, it may be accredited to the fact that hormonal behavior under cold stress conditions is inclined by cross-talk with signaling forces the consulting response to other environmental temptations including light [39] and is also obstructed by endogenous developing programs, especially those, which results in developmental phase transitions [40].

## 5. Response of plant hormones to drought stress

Although the green revolution has enhanced plant output, the ever-increasing global population and global warming (which is producing drought stress) are again putting a strain on our ability to feed the globe [41]. As a result of catastrophic losses in agricultural output caused by drought stress during the previous few decades, there is a worldwide challenge to enhance the yield and plant drought resistance [42]. Drought is a disaster for agriculture, humankind, and animals alike. Climate change is bringing us closer to a hotter and drier planet. Hence, we urgently need to produce drought-resistant, high-yielding crops [43]. Drought is an important plant stressor that considerably influences plant growth and productivity, resulting in large agricultural production losses [44]. Drought stress causes plants to undergo morphological, physiological, biochemical, and molecular changes [45].

Small compounds such as peptides or hormone are effective in agriculture for fine-tuning drought response pathways while maintaining production. Research initiatives that reveal the physiology of plant responses to drought in model systems and transfer these results to crops will result in novel water-saving measures [46]. Even as the world's population rises, finding solutions to alleviate agriculture's "thirst" would lessen competition for freshwater supplies [47]. Exogenous application of hormones improves the endogenous hormone contents that significantly help in improved photosynthetic fluorescence of leaves, plant enzymatic activities, regulates source-sink balance, yields maintenance, and enhances carbon metabolism under the drought stress environment [48].

To reduce the effect of drought stress on plants, these mechanisms may include the mitogen-activated protein kinases (MAPK) signaling system, calcium signaling route, transcription factor modulation, and higher levels of antioxidant enzymes and other chemicals [49]. Exogenous application of chemicals (nitric oxide, 24-epibrassinolide, proline, glycine betaine), plant breeding, and transgenic approaches is under consideration by scientists to enhance these systems [50]. Early-stage application helps to control drought stress, and spraying 6-BA has capability to regulate the content of endogenous hormones and improve photosynthetic characteristics of sweet potato, and thus effectively alleviates the loss of yield [51]. Under drought condition, seed treatment of plant hormones, maize germination percentage, and seedling growth were enhanced significantly [52] amid to improved drought resistance.

## 6. Involvement of phytohormones in resetting plant-pest interaction

Many aspects of plant growth and responses to biotic and abiotic stressors are influenced by phytohormones [53]. Jasmonates, ethylene, and salicylic acid are the three primary phytohormones that mediate defensive responses to pests and diseases. Other hormones' involvement in defensive signaling has recently been discovered. Abscisic acid, a hormone generally linked with abiotic stress reactions, has been identified as a critical fine-tune regulator of defenses [54].

Plants have their defense systems to combat pests' invasions. It has limited pre-existing defense systems, and the majority of the defense response is triggered only after an insect or disease has infected it. Pests' attacks are repelled both directly and indirectly by plants. Many of these defenses are controlled by signaling pathways in which phytohormones play a significant role. Insects evolve methods to overcome the plant obstacles simultaneously, resulting in an intriguing co-evolution of plant-pest interactions. Biological immune systems require the ability to sense and respond, but to what degree can plants recognize and respond to pests specifically is integral? The jasmonate route has emerged as a critical signaling mechanism for integrating data from the plant-pest contact into broad-spectrum defensive responses [55]. A proper defense reaction to a biotic danger requires early detection. Pathogens are identified when pattern recognition receptors (PRRs) on the surface of the host plant cell detects conserved patterns of microbial molecules termed microbe- or pathogen-associated molecular patterns (MAMPs or PAMPs), resulting in PAMP-triggered immunity (PTI). PRRs also identify damage-associated molecular patterns (DAMPs) that are endogenous chemicals generated by the plant after infection and induce defensive responses [56]. Pathogens can avoid this innate immune response by using effector proteins that decrease PTI when delivered into the host cell. Disease resistance proteins identified in some plant genotypes precisely detect pathogen effectors, resulting in effector-triggered immunity (ETI) [57]. However, both PTI and ETI models are considered as a generalization [58]. Involved receptors and ligands in molecular identification have empowered conclusions about the specificity of recognition in plant-pathogen interactions. Usually, ETI is activated by highly pathogen-specific chemicals, whereas PTI is based on the non-specific detection of common microbial molecules [59].

In contrast to diseases, insects are vastly complex multicellular animals with a wide range of lives and cognitive behaviors. The plant may use cues from these patterns to identify the threat of herbivory and establish appropriate defense responses [60]. When plant tissue integrity is disrupted during insect feeding, jasmonoyl-L-isoleucine (JA-Ile) is produced, and a well-defined signal transduction chain is activated, leading to the transcriptional activation of defensive responses [61]. Beyond jasmonates, how can plants fine-tune their defensive mechanism to mount herbivore-specific responses? There are two possible responses to this topic. The first is that plants may employ jasmonates-independent, parallel pathways to generate unique response patterns. Second, specificity may be mediated by the activation of jasmonates-response spatiotemporal modulators. The first premise is supported by research on the tomato's recognition and response mechanism to the potato aphid *Macrosiphum euphorbiae*. *Mi-1*, a potential receptor, causes salicylic acid-mediated signaling [62] and resistance independent of the jasmonate system [63]. Plant recognition and response to many different hemipterans appear to follow a similar pattern, implying that plants utilize jasmonates separate hormone response mechanisms to develop specialized resistance to phloem-feeding [64].

The capacity of one participant to notice and respond to cues provided by the other is a recurring topic in all realms of plant-pest biology. This information flow serves as a great focal point for revealing fundamental chemical and molecular principles of plant-pest interactions. Nonetheless, the research supports several broad conclusions concerning plant-pest interactions' distinctiveness. To begin, plants recognize distinct arthropods by combining various environmental signals, ranging from mechanical stimulation by insects moving on plant surfaces to contact with glandular components during feeding. Second, herbivore sensing activates regulatory responses involving several phytohormones, with the Jasmonate pathway playing a pivotal role in host resistance. Third, despite the significant conservation of Jasmonate signaling, it is becoming increasingly evident that several hormone response pathways interact to transform initial sensory events into suitable responses that promote plant fitness in the face of aggressive aggressors. Anyhow, understanding the impacts of phytohormones at the whole plant level is undoubtedly important, and the future study presents an intriguing challenge.

## 7. Future prospects and challenges

Plant hormones have proven themselves effective against biotic and abiotic stresses that emerged due to climate change. However, it is a challenge for researchers and industrialists to prepare cost-effective products that can be used on a large scale. Moreover, salicylic acid is a defensive hormone, but its biosynthesis pathway for salicylic acid is still incomplete; thus, this section requires attention. Furthermore, various parts of the globe face more biotic/abiotic stress, but we still lack understanding of the response of jasmonates against multiple stress conditions. In addition, optimizing the endogenous levels of plant hormones to improve the stress-responsive crosstalk mechanisms between multiple hormones is still a research gap.

## 8. Conclusion

The increasing population requires the scientific community to suggest effective technologies for food security. Besides this, climate change is a global problem that adversely impacts the agriculture sector and creates hurdles for agriculture sustainability. Plant hormones within the plants or prepared by the plants have a promising role in tackling these challenges. It enables the plants to tolerate the biotic and abiotic stresses as well as improve agriculture productivity. Therefore, understanding the mechanism and endogenous application of plant hormones can be an effective tool against climate change and its driven problems.

## Conflict of interest

The authors have no conflict of interests to declare.

## Author details

Muhammad Amjad Bashir[1*†], Qurat-Ul-Ain Raza[2†], Hafiz Muhammad Ali Raza[1,2], Muhammad Umair Sial[3], Abdur Rehim[1,2*], Kashif Ali Khan[1], Muhammad Ijaz[1] and Muhammad Wasif[1]

1 College of Agriculture, Bahadur Sub-Campus Layyah, Bahauddin Zakariya University, Multan, Pakistan

2 Department of Soil Science, Bahauddin Zakariya University, Multan, Pakistan

3 Department of Entomology, University of Agriculture, Faisalabad, Pakistan

*Address all correspondence to: amjad.bashir941@gmail.com and abdur.rehim@bzu.edu.pk

† The authors have contributed equally, should be considered as the first authors.

## References


[1] Arora NK. Impact of climate change on agriculture production and its sustainable solutions. Environmental Sustainability. 2019;**2**:95-96. DOI: 10.1007/s42398-019-00078-w

[2] Aryal JP, Sapkota TB, Khurana R, Khatri-Chhetri A, Rahut DB, Jat ML. Climate change and agriculture in South Asia: Adaptation options in smallholder production systems. Environment, Development and Sustainability. 2020;**22**:5045-5075. DOI: 10.1007/s10668-019-00414-4

[3] Malhi GS, Kaur M, Kaushik P. Impact of climate change on agriculture and its mitigation strategies: A review. Sustainability. 2021;**13**:1-21

[4] Karavolias NG, Horner W, Abugu MN, Evanega SN. Application of gene editing for climate change in agriculture. Frontiers in Sustainable Food Systems. 2021;**5**:296

[5] Du F, Ruan G, Liu H. Analytical methods for tracing plant hormones. Analytical and Bioanalytical Chemistry. 2012;**403**:55-74. DOI: 10.1007/s00216-011-5623-x

[6] Rehim A, Amjad Bashir M, Raza Q-U-A, Gallagher K, Berlyn GP. Yield enhancement of biostimulants, vitamin B12, and CoQ10 compared to inorganic fertilizer in radish. Agronomy. 2021;**11**:697. DOI: 10.3390/agronomy11040697

[7] Bashir MA, Rehim A, Raza Q-U-A, Muhammad Ali Raza H, Zhai L, Liu H, et al. Biostimulants as plant growth stimulators in modernized agriculture and environmental sustainability. In: Ahmad F, Sultan M, editors. Technology in Agriculture. London, UK: IntechOpen; 2021

[8] Peleg Z, Blumwald E. Hormone balance and abiotic stress tolerance in crop plants. Current Opinion in Plant Biology. 2011;**14**:290-295

[9] Shawky Soliman A. Plant growth hormones. In: Cell Growth. London, UK: IntechOpen; 2020

[10] Siddiqi KS, Husen A. Plant response to jasmonates: Current developments and their role in changing environment. Bulletin of the National Research Centre. 2019;**43**:1-11. DOI: 10.1186/s42269-019-0195-6

[11] Sun LR, Wang YB, He SB, Hao FS. Mechanisms for abscisic acid inhibition of primary root growth. Plant Signaling & Behavior. 2018;**13**:e1500069

[12] Schaller GE, Street IH, Kieber JJ. Cytokinin and the cell cycle. Current Opinion in Plant Biology. 2014;**21**:7-15

[13] Iqbal N, Khan NA, Ferrante A, Trivellini A, Francini A, Khan MIR. Ethylene role in plant growth, development and senescence: Interaction with other phytohormones. Frontiers in Plant Science. 2017;**8**:475. DOI: 10.3389/fpls.2017.00475

[14] Ribba T, Garrido-Vargas F, O'Brien JA. Auxin-mediated responses under salt stress: From developmental regulation to biotechnological applications. Journal of Experimental Botany. 2020;**71**:3843-3853

[15] Yoshida H, Ueguchi-Tanaka M, Matsuoka M. Regulatory networks acted upon by the GID1-DELLA system after perceiving gibberellin. In: Enzymes. Vol. 35. Massachusetts, USA: Academic Press; 2014. pp. 1-25

[16] Hassoon AS, Abdulsattar Abduljabbar I. Review on the role of salicylic acid in plants. In: Sustainable Crop Production. London, UK: IntechOpen; 2020

[17] Fariduddin Q, Yusuf M, Ahmad I, Ahmad A. Brassinosteroids and their

role in response of plants to abiotic stresses. Biologia Plantarum. 2014;**58**:9-17

[18] Qureshi MA, Jaskani MJ, Khan AS, Ahmad R. Influence of endogenous plant hormones on physiological and growth attributes of Kinnow mandarin grafted on nine rootstocks. Journal of Plant Growth Regulation. 2021:1-11. DOI: 10.1007/s00344-021-10380-9

[19] Akhtar SS, Mekureyaw MF, Pandey C, Roitsch T. Role of cytokinins for interactions of plants with microbial pathogens and pest insects. Frontiers in Plant Science. 2020;**10**:1777

[20] Emenecker RJ, Strader LC. Auxin-abscisic acid interactions in plant growth and development. Biomolecules. 2020;**10**:281. DOI: 10.3390/BIOM10020281

[21] Ghorbel M, Brini F, Sharma A, Landi M. Role of jasmonic acid in plants: The molecular point of view. Plant Cell Reports. 2021;**40**:1471-1494

[22] Binenbaum J, Weinstain R, Shani E. Gibberellin localization and transport in plants. Trends in Plant Science. 2018;**23**:410-421

[23] Ahammed GJ, Li X, Liu A, Chen S. Brassinosteroids in plant tolerance to abiotic stress. Journal of Plant Growth Regulation. 2020;**39**:1451-1464

[24] Ahanger MA, Ashraf M, Bajguz A, Ahmad P. Brassinosteroids regulate growth in plants under stressful environments and crosstalk with other potential phytohormones. Journal of Plant Growth Regulation. 2018; **37**:1007-1024

[25] Hussain S, Peng S, Fahad S, Khaliq A, Huang J, Cui K, et al. Rice management interventions to mitigate greenhouse gas emissions: A review. Environmental Science and Pollution Research. 2015;**22**:3342-3360. DOI: 10.1007/s11356-014-3760-4

[26] Fahad S, Hussain S, Matloob A, Khan FA, Khaliq A, Saud S, et al. Phytohormones and plant responses to salinity stress: A review. Plant Growth Regulation. 2015;**75**:391-404

[27] Yang Y, Guo Y. Unraveling salt stress signaling in plants. Journal of Integrative Plant Biology. 2018; **60**:796-804

[28] Ouhibi C, Attia H, Rebah F, Msilini N, Chebbi M, Aarrouf J, et al. Salt stress mitigation by seed priming with UV-C in lettuce plants: Growth, antioxidant activity and phenolic compounds. Plant Physiology and Biochemistry. 2014;**83**:126-133. DOI: 10.1016/j.plaphy.2014.07.019

[29] Zhu JK. Salt and drought stress signal transduction in plants. Annual Review of Plant Biology. 2002; **53**:247-273

[30] Yang G, Yu Z, Gao L, Zheng C. SnRK2s at the crossroads of growth and stress responses. Trends in Plant Science. 2019;**24**:672-676

[31] Ku YS, Sintaha M, Cheung MY, Lam HM. Plant hormone signaling crosstalks between biotic and abiotic stress responses. International Journal of Molecular Sciences. 2018;**19**:3206

[32] Davies PJ. Plant Hormones: Biosynthesis, Signal Transduction, Action! Netherlands: Springer; 2010

[33] Skirycz A, Inzé D. More from less: Plant growth under limited water. Current Opinion in Biotechnology. 2010;**21**:197-203

[34] Harrison MA. Cross-talk between phytohormone signaling pathways under both optimal and stressful environmental conditions. In: Phytohormones and Abiotic Stress Tolerance in Plants. Berlin Heidelberg: Springer-Verlag; 2012. pp. 49-76

[35] Ruelland E, Vaultier MN, Zachowski A, Hurry V. Cold signalling and cold acclimation in plants. Advances in Botanical Research. 2009;**49**:35-150

[36] Bahuguna RN, Jagadish KSV. Temperature regulation of plant phenological development. Environmental and Experimental Botany. 2015;**111**:83-90. DOI: 10.1016/j.envexpbot.2014.10.007

[37] Mittler R, Finka A, Goloubinoff P. How do plants feel the heat? Trends in Biochemical Sciences. 2012;**37**:118-125

[38] Santner A, Calderon-Villalobos LIA, Estelle M. Plant hormones are versatile chemical regulators of plant growth. Nature Chemical Biology. 2009;**5**:301-307

[39] Franklin KA. Light and temperature signal crosstalk in plant development. Current Opinion in Plant Biology. 2009;**12**:63-68

[40] Achard P, Cheng H, De Grauwe L, Decat J, Schoutteten H, Moritz T, et al. Integration of plant responses to environmentally activated phytohormonal signals. Science. 2006; **311**:91-94. DOI: 10.1126/science.1118642

[41] Godfray HCJ, Beddington JR, Crute IR, Haddad L, Lawrence D, Muir JF, et al. Food security: The challenge of feeding 9 billion people. Science. 2010;**327**:812-818

[42] Naveed M, Mitter B, Reichenauer TG, Wieczorek K, Sessitsch A. Increased drought stress resilience of maize through endophytic colonization by *Burkholderia phytofirmans* PsJN and *Enterobacter* sp. FD17. Environmental and Experimental Botany. 2014;**97**:30-39

[43] Seleiman MF, Al-Suhaibani N, Ali N, Akmal M, Alotaibi M, Refay Y, et al. Drought stress impacts on plants and different approaches to alleviate its adverse effects. Plants. 2021;**10**:1-25

[44] Zia R, Nawaz MS, Siddique MJ, Hakim S, Imran A. Plant survival under drought stress: Implications, adaptive responses, and integrated rhizosphere management strategy for stress mitigation. Microbiological Research. 2021;**242**:126626

[45] Tardieu F, Simonneau T, Muller B. The physiological basis of drought tolerance in crop plants: A scenario-dependent probabilistic approach. Annual Review of Plant Biology. 2018;**69**:733-759

[46] Daszkowska-Golec A. The role of abscisic acid in drought stress: How ABA helps plants to cope with drought stress. In: Drought Stress Tolerance in Plants, Vol 2: Molecular and Genetic Perspectives. New York City, USA: Springer International Publishing; 2016. pp. 123-151

[47] Jury WA, Vaux HJ. The emerging global water crisis: Managing scarcity and conflict between water users. Advances in Agronomy. 2007;**95**:1-76

[48] Huan L, Jin-qiang W, Qing L. Photosynthesis product allocation and yield in sweet potato with spraying exogenous hormones under drought stress. Journal of Plant Physiology. 2020;**253**:153265. DOI: 10.1016/J.JPLPH.2020.153265

[49] Forni C, Duca D, Glick BR. Mechanisms of plant response to salt and drought stress and their alteration by rhizobacteria. Plant and Soil. 2017;**410**:335-356. DOI: 10.1007/s11104-016-3007-x

[50] Liaqat S, Umar S, Saffeullah P, Iqbal N, Siddiqi TO, Khan MIR. Protective effect of 24-epibrassinolide on barley plants growing under combined stress of salinity and potassium deficiency. Journal of Plant Growth Regulation. 2020;**39**:1543-1558. DOI: 10.1007/s00344-020-10163-8

[51] Wang JQ, Li H, Liu Q, Zeng LS. Effects of exogenous plant hormones on

physiological characteristics and yield of sweet potato under drought stress. Ying Yong Sheng tai xue bao = The Journal of Applied Ecology. 2020;**31**:189-198. DOI: 10.13287/J.1001-9332.202001.026

[52] Yuan Z, Wang C, Li SP, Li X, Tai F. Effects of different plant hormones or PEG seed soaking on maize resistance to drought stress. Canadian Journal of Plant Science. 2014;**94**:1491-1499. DOI: 10.4141/CJPS-2014-110/ASSET/IMAGES/LARGE/CJPS-2014-110F6.JPEG

[53] Pieterse CMJ, Leon-Reyes A, Van Der Ent S, Van Wees SCM. Networking by small-molecule hormones in plant immunity. Nature Chemical Biology. 2009;**5**:308-316

[54] Asselbergh B, De Vleesschauwer D, Höfte M. Global switches and fine-tuning-ABA modulates plant pathogen defense. Molecular Plant-Microbe Interactions. 2008;**21**:709-719

[55] Stout MJ, Thaler JS, Thomma BPHJ. Plant-mediated interactions between pathogenic microorganisms and herbivorous arthropods. Annual Review of Entomology. 2006;**51**:663-689

[56] Boller T, Felix G. A renaissance of elicitors: Perception of microbe-associated molecular patterns and danger signals by pattern-recognition receptors. Annual Review of Plant Biology. 2009;**60**:379-407. DOI: 10.1146/annurev.arplant.57.032905.105346

[57] Jones JDG, Dangl JL. The plant immune system. Nature. 2006; **444**:323-329

[58] Thomma BPHJ, Nürnberger T, Joosten MHAJ. Of PAMPs and effectors: The blurred PTI-ETI dichotomy. Plant Cell. 2011;**23**:4-15. DOI: 10.1105/tpc.110.082602

[59] Dodds PN, Lawrence GJ, Catanzariti AM, Teh T, Wang C-IIA, Ayliffe MA, et al. Direct protein interaction underlies gene-for-gene specificity and coevolution of the flax resistance genes and flax rust avirulence genes. Proceedings of the National Academy of Sciences of the United States of America. 2006;**103**:8888-8893. DOI: 10.1073/pnas.0602577103

[60] Hilker M, Meiners T. How do plants "notice" attack by herbivorous arthropods? Biological Reviews. 2010;**85**:267-280. DOI: 10.1111/j.1469-185X.2009.00100.x

[61] Heil M. Damaged-self recognition in plant herbivore defence. Trends in Plant Science. 2009;**14**:356-363. DOI: 10.1016/j.tplants.2009.04.002

[62] Li Q, Xie QG, Smith-Becker J, Navarre DA, Kaloshian I. Mi-1-mediated aphid resistance involves salicylic acid and mitogen-activated protein kinase signaling cascades. Molecular Plant-Microbe Interactions. 2006;**19**:655-664. DOI: 10.1094/MPMI-19-0655

[63] Bhattarai KK, Xie QG, Pourshalimi D, Younglove T, Kaloshian I. Coi1-dependent signaling pathway is not required for Mi-1-mediated potato aphid resistance. Molecular Plant-Microbe Interactions. 2007;**20**:276-282. DOI: 10.1094/MPMI-20-3-0276

[64] Dogimont C, Bendahmane A, Chovelon V, Boissot N. Host plant resistance to aphids in cultivated crops: Genetic and molecular bases, and interactions with aphid populations. Comptes Rendus-Biologies. 2010;**333**:566-573. DOI: 10.1016/j.crvi.2010.04.003

Chapter 6

# Seed Priming with Phytohormones

*Musa Saheed Ibrahim, Nathan Moses and Beckley Ikhajiagbe*

## Abstract

Improving growth and yield properties of plats has been the major aim of most researchers in plat science field. Several strategies have been suggested in order to sustainably improve crop yield. Among these strategies is biopriming, has gained the highest attention being the most effective strategy. Biopriming is a technique involving pre-soaking of plant seed into a solution in order for the metabolic processes to be enhanced before to germination, thereby improving the percentage and rate of germination and increase seedling growth and crop yield under normal and different environmental stresses. The most important aspects of phytohormones is that they are very essential in the regulation of plant development and growth and also functions as an essential chemical messengers, allowing plants to thrive even during exposure to various stresses. Priming plant seeds with phytohormones has led to improved growth and yield of plants in developing countries. Furthermore, it has emerged as an important tool for mitigating the effects of environmental stress. However, this innovation has received less attention from local farmers and merger work has been reported. Therefore, this review discusses the mechanism and potential role of priming with phytohormones to enhance crop productivity and improve plant tolerance to biotic and abiotic stressors.

**Keywords:** biopriming, phytohormones, plant growth regulators, biotic stress, abiotic stress

## 1. Introduction

Plant hormones since their discovery in seventeenth century have been used extensively in crop production. Advent of technology allowed the researchers to study more about different plant hormones and their endogenous and exogenous uses [1]. Plant hormones are a group of naturally occurring organic substances that are produced by plants and has effects on plant physiological processes when released at low concentrations. Plant hormones also referred to as phytohormones have the ability to influence growth, differentiation, development, and stomatal movement [2]. These hormones are well known for their important roles in plant physiology such as regulation of plant growth and development and important chemical messenger [3]. The first plant hormone that was identified was auxin. This plant hormone has a wide range effect on plant growth and development. Recently, it has been clearly shown that plant hormones are not exactly like the mammalian hormones. Even though the synthesis of plant hormone may be localized as in the case of animal hormones, but plant hormones may be transported to a long distance where it is needed most [4]. Auxins (IAAs), cytokinins (CKs), gibberellins (GAs),



IntechOpen

abscisic acid (ABA), salicylic acid (SA), and ethylene (ET) are of the most essential phytohormones that are important for plant growth and development [5, 6]. These plant hormones have series of ways by which they improve growth and development of plants.

Strategies and mechanisms of growth promotion by phytohormones have been assessed, and several hormones have been categorized in different classes [7]. Major functions of these phytohormones are cell enlargement such as in auxin [6]. Cell division such as auxin, which stimulates the division of cells in the cambium and sometimes together with cytokinin in tissue culture media [3]. Vascular tissue differentiation such as in indole acetic acid which stimulates fast differentiation of phloem and xylem, initiation of plant root, and also the development of secondary roots under normal growth media and tissue culture media [6]. In most cases for plants, synthesis of hormones and plant response determines plant health and may in turn improve soil fertility [8]. Plant hormones have been very helpful in sustainable agriculture by stimulate fast growth and development in plants, which can be achieved through processes such as by spraying on the leaf and other plant tissues [7]. It is now important to consider introducing the endogenous and exogenous plant hormones into the growing seeds to improve stimulating rates. This process is called seed biopriming.

Seed biopriming is a unique innovation where seeds are treated with biological substances (such as bacteria, fungi, and hormones) that assimilates plant morphology and physiological facets [9]. This process has also been used to fortify plant against diseases [10]. Hormonal priming is a situation where plant seed are prime with phytohormones. This process has been documented to influence seed metabolism and germination rate. This technique has now been adopted in developing countries to enhance seed germination, growth of seedlings, and crop yield in environmentally disturbed arid lands [11, 12]. Ensuring better germination, improved plant growth and seedling vigor through seed biopriming would result in healthy and productive plants. There are several approaches to seed priming, however, all followed similar mechanisms and are used in improving plant growth properties [8]. Therefore, the purpose of this review is to summarize the effectiveness of seed priming using phytohormones in enhancing crop productivity along with future prospects of this innovative technology. In order to achieve this, this review discusses mechanisms involved in hormonal seed priming, hormone specific in seed priming, biopriming and crop productivity, role of hormonal priming in plant stress mitigation, economics of hormonal priming and future prospects of hormonal priming.

## 2. Mechanisms involved in seed biopriming

There are several approaches used in biopriming such as hydro priming which involve soaking seeds in distilled water and oven drying at low temperature before sowing. This process does not involve the use synthetic chemicals which makes it faster, cheaper, and eco-friendlier [13]. Osmo-priming has also been considered by local farmers in Brazil where seeds are soaked in salt-containing solution. This process allows slow imbibition of water into the seed and that initiate energy activation [14]. Hormopriming which involves the soaking of seeds in naturally occurring plant growth regulators. This process has direct effect on processes of seed metabolism. Usually, scientist considered abscisic acid, auxins, gibberellins, kinetin, ethylene, polyamines, and salicylic acid as the hormones widely used in priming [15]. For example, Galhaut et al. [16] confirmed the effectiveness of gibberellic acid in improving photosynthetic properties, antioxidant system, seedling

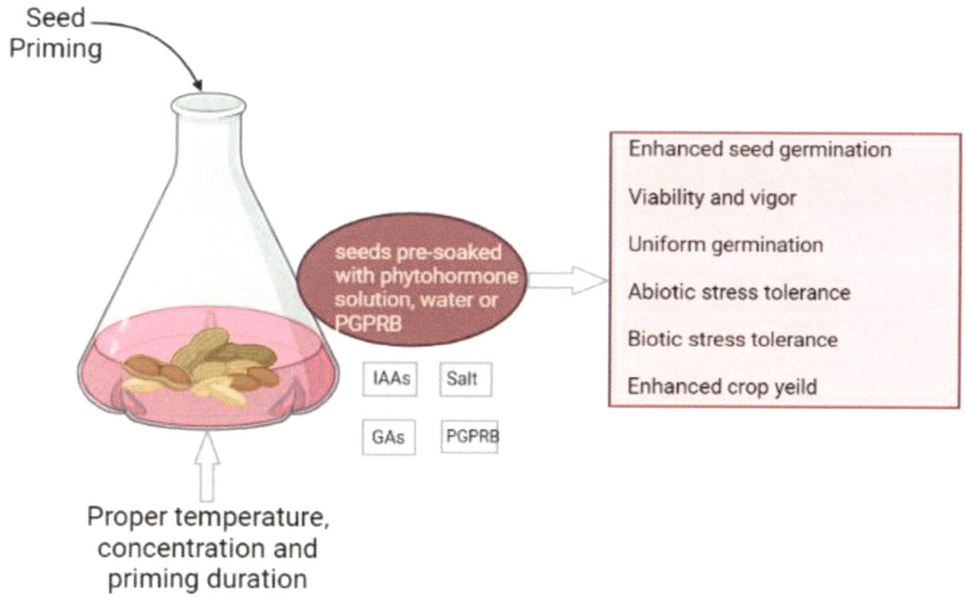


**Figure 1.**
*Schematic model indicating the strategy of seed priming and the possible results.*

emergence, and growth in white clover plant grown on heavy metal polluted soil. Other strategies such as chemo-priming [17], nutri-priming [18], and plant growth promoting bacteria rhizobacteria (PGPRB) priming [19, 20] have proved effective in promoting growth properties of plants.

Generally, mechanism involve in seed biopriming include pre-soaking of seeds with a particular concentration of priming agent such as water, PGPRB, or phytohormone (**Figure 1**). This process improves germination parameters, seedling yield and growth, by either increasing nutrient utilization with the help of an improved physiological activities and root cell differentiation and division [21, 22], or by stimulating the activation of important metabolites such as amylase which initiate energy supply and improve germination properties [17]. Previous researches have documented evidence on seed priming with phytohormones in varieties of plant species and how important physiological processes such as growth and development, respiration, and transpiration are improved [23, 24]. The results of these researches have shown the significant roles of phytohormones in the biochemical, defense, and signaling pathways of plants [6]. Many researchers are now working to develop effective approaches to alleviate biotic and abiotic stresses and enhance crop production, especially as the world is constantly facing global warming. Seed priming with phytohormones can modulate the physiological and genetic mechanisms, making plants capable of tolerating these environmental stresses or making plants resistance to the stressors. These mechanisms if well adopted would be promising.

## 3. Hormone specific in seed biopriming

There are various plant hormones produced by plants which have various functions. However, auxins, cytokinins, gibberellins, abscisic acid, salicylic acid, and ethylene are the most frequently used in seed priming. In addition, methyl jasmonate have also been used by previous literature in seed priming. According to

| S/N | Plant | Abiotic stressor | Phytohormone | Response | Reference |
|---|---|---|---|---|---|
| 1. | Rice (*Oryza sativa*) | Salt | Auxin | Modulate ion homeostasis in rice plant | Iqbal and Ashraf [25] |
| 2. | Soybean (*Glycine max*) | Drought | Cytokinin | Enhanced drought tolerance in soybean plants | Mangena [26] |
| 3. | Wheat (*Triticum aestivum*) | Salt | Cytokinin | Decreased electrolyte leakage and improved salt tolerance | Angrish et al. [27] |
| 4. | Pot marigold and sweet fennel | Salt | Gibberellin | Improved dry matter and increased tolerance to salinity by enhancing antioxidant enzyme activities | Sedghi et al. [28] |
| 5. | Sorghum (*Sorghum bicolor*) | Drought | Gibberellin | Increased CAT and APX activities | Sheykhbaglou et al. [29] |
| 6. | Rice (*O. sativa*) | Flood | Gibberellin | Increased α-amylase activity, sucrose, glucose, and fructose content in seeds. | Watanabe et al. [30] |
| 7. | Maize (*Zea mays*) | Salt | Gibberellin | Reduced negative effect of salt stress | Hamza and Ali [31] |
| 8. | Wheat (*T. aestivum*) | Salt | Abscisic acid | Increased α-amylase activity, sucrose, glucose, and fructose content in seeds. | Zongshuai et al. [32] |
| 9. | *Agropyron elongatum* | Temperature | Abscisic acid | Enhanced anti-oxidant enzyme activities | Gurmani et al. [33] |
| 10. | Rice (*O. sativa*) | Chromium | Salicylic acid | Enhanced antioxidant enzyme activities, detoxified ROS | Pouramir-Dashtmian et al. [34] |
| 11. | Maize (*Z. mays*) | Chilling | Salicylic acid | Enhanced enzymatic antioxidant activities, high tissue water content | Farooq et al. [35] |
| 12. | Lucerne (*Medicago sativa* L.) | Cadmium | Jasmonic acid | Increased α-amylase activity and sugar content | Watanabe et al. [30] |

| S/N | Plant | Abiotic stressor | Phytohormone | Response | Reference |
|---|---|---|---|---|---|
| 13. | Rice (*O. sativa*) | Salt | Ethylene | Increased α-amylase activity and sugar content | Dai et al. [36] |
| 14. | Sorghum (*S. bicolor* L.) | Iron toxic soil with high pH | Indole acetic acid, gibberellin, and ascorbic acid | Improve seed germination even under elevated pH levels | Onoabhagbe et al. [37] |
| 15. | Sorghum (*S. bicolor* L.) | Iron toxic soil | Indole acetic acid, gibberellin, and ascorbic acid | Improved germinability and germination properties | Beckley et al. [38] |

**Table 1.**
*Seed priming with phytohormones for developing abiotic stress tolerance in plants.*

several plant hormones function in seed germination and energy production for the developing embryo.

### 3.1 Gibberellin

Gibberellins (GAs) are plant hormones that have shown the capacity to regulate different developmental processes, such as stem elongation, germination, dormancy, flowering initiation, flower development, fruit development, and leaf and fruit senescence [3]. Gibberellin brings results that are somewhat similar to the ones by auxin, even though their mode of action differs [21].

Furthermore, gibberellin plays an essential function various physiological and developmental stages in plants (**Table 1**), but they are more effective in making stems increasing stem elongation through rapid cell differentiation and cell circle [3], Thereby leading to internode elongation. Dwarf and rosette plants (plants with little space between nodes on a stem or plants with clustered base) have been investigated to have low or no concentration of gibberellin. Also, Onoabhagbe et al. [37] used gibberellin and other plant hormones to improve germination properties and growth of *Sorghum bicolor* under elevated pH regime using chemoprming system.

### 3.2 Auxin

Auxin (IAAs) were also one of the most essential and the first identified phytohormone. IAA phytohormone is known to show an essential role in modulating plant growth and developmental processes, especially the root growth, cell elongation, vascular differentiation, and apical dominance [3]. IAAs also play an important function in cell division and differentiation, in fruit developmental stages, in the root formation from cuttings, in the lateral branching (apical dominance) inhibition, and in the leaf fall (abscission) frequencies. IAA conjugates is usually the form assumed by IAA in higher plants, IAAs conjugates functions as the primary free endogenous auxin that brings about plant developmental processes. The exogenous priming of seeds with IAAs induces the fast and improved formation of adventitious roots and lateral roots [39].

One of the most important naturally existing auxin is β-indolylacetic acid (IAA), which is obtained either from the amino acid tryptophan or from the breakdown of carbohydrates known as glycosides. This chemical influence plants by its activity on the chemical bonds linking carbohydrates present in plant cell walls. The cycle allows the cells to be irreversibly adjusted and is joined by the passage of water and the synthesis of new cell wall material.

Auxin is engaged with cell development and cell extension of certain parts of a developing plant such as the stem which produced basically in pieces of the plant that are effectively developing like the stem (specifically, the stem tip). The phototropic reaction happens on the grounds that more amounts of auxin are disseminated to the side away from the light than to the side toward it, making the concealed side stretch all the more firmly and accordingly bend the stem toward the light. Additionally, the geotropic reaction happens in light of the fact that more auxin gathers along the lower side of the developing stem than along the upper side, creating a vertical arch.

### 3.3 Cytokinins

Cytokinins (CKs) are one of the plant hormones that are known to regulate various sections of plant growth and development, such as cell division, apical development, root elongation, stomatal behavior, and chloroplast synthesis [26]. It has been widely documented that application of CKs can promote crop development and yield. For example, Fricke et al. [40] demonstrated the use of CKs to improve cotton seedlings development. The result showed an increase in cotton yield of 10%. Another important aspect of CKs is its ability to improve pathogenesis in plant. Furthermore, CKs application has showed resistance against *Pseudomonas syringae* in *Arabidopsis thaliana* [28] and *Nicotiana tabacum* [31]. CKs are majorly produced in the root regions from a compound known as adenine. They are found moving upward within xylem (woody tissue) and then pass it to the leaves and fruits, where they are needed for growth and cell differentiation in plants.

CKs also functions together with auxin to reverse senescence in plants through stabilizing protein levels and synthesis of chlorophyll in the leaf. Senescence is a developmental stage in plant when the yellowing of leaves is visible as a result of protein breakdown and chlorophyll is decomposition. CKs also can also be used in the storage of green vegetables to reduce yellowing [7]. In horticultural tissue culture, according to Addicott [41], increased auxin and reduced cytokinin conditions can lead to improved root development, while reduced auxin and increased cytokinin conditions can lead to improved shoot development.

### 3.4 Ethylene

Ethylene (ET) is another essential plant hormone that influence ripening and rotting of fruit in plants [42]. ET is a very important plant hormone because it is the only plant hormone occurring as a gas. Furthermore, ET can be synthesized in almost every part of a plant, and can diffuse as a gas through the plant's tissue, outside the plant, and travel through the air to affect other plants within the vicinity. For example, Montalvo et al. [43] reported accelerated mango ripening as a result of application of ET for 12 hours. This process stimulating process was achieved through the production of 1-amino cyclopropane-1-carboxylic acid (ACC: an ethylene precursor) and improved ACC oxidase activity.

### 3.5 Abscisic acid

Abscisic acid (ABA) is another essential plant hormone that is known to stimulate developmental processes in plants, such as bud elongation, dormancy, control of organ size, and stomatal closure. It is also known as stress hormone because it plays essential function in regulating plant responses to various biotic and abiotic environmental stressors such as drought, salinity, cold, heavy metals stress, and heat stress. ABA is a stress-triggered hormone, such that the highest concentration of ABA is synthesized in the root region of plant in response to decreased soil water potential (which is associated with dry soil) and other stress induced conditions. After the synthesis, ABA is then translocated to the leaves regions, where it gradually affect the osmotic potential of stomatal guard cells, causing them to shrink, leading to the closure of stomata. The ABA-induced stomatal closure brings about reduced transpiration (evaporation of water out of the stomata), thus preventing further water loss from the leaves in times of low water availability. A close linear correlation was found between the ABA content of the leaves and their conductance (stomatal resistance) on a leaf area basis.

### 3.6 Salicylic acid

Salicylic acid (SA) is also an essential plant hormone, belonging to the phenolic group. It has various physiological benefits in plants because of its ability to regulate the processes of growth and development in plants such as photosynthesis, respiration, transpiration, and the transportation of ions in plants. According to Devinar et al. [44], Khan et al. [45], Senaratna et al. [46], and Bastam et al. [47], SA exhibits an essential role in the activation, modulation, and regulation of numerous responses during plant exposure to abiotic and biotic environmental stresses. Fahad and Bano [39] reported that SA has the capabilities to initiate and generate a cascade of several signaling pathways by interacting with other plant hormones such as ABA and ET and plays an important role in mitigating plant stresses. Ikhajiagbe and Musa [15] investigated the effect of SA on germination and early seedling growth of pigeon pea (*Cajanus cajan*). The research showed that increased levels of SA at 20 mg/L is very essential for maximum seed germination and early seedling growth of *C. cajan*. Furthermore, Jadhav and Bhamburdekar [48] observed the positive influence of SA on the root and shoot growth of groundnut. Anaya et al. [49] indicated the significant contribution of SA in alleviating saline stress in *Vicia faba* under salt stress condition.

## 4. Bio-priming with phytohormones and crop productivity

According to Khan [50], seed priming involves various physiological treatments that can improve seed germination and seedling vigor through the addition of effective plant hormones. Ikhajiagbe and Musa [15] reported that pigeon pea (*C. cajan*) seeds bioprimed with salicylic acid for 20 hours results in improved germination properties. Malathi and Doraisamy [51] observed that seed priming with gibberellins protected seeds of groundnut from the infection of *Macrophomina phaseolina* and also bring about improved seedling vigor, plant dry matter, and prevented loss of oil content for up to 6 months of storage. Mohamedy et al. [52] discovered that biopriming of pea seeds with ethylene showed significant decrease in pre-emergence damping off, that is occasioned as a result of infestation by *Fusarium solani* in abandoned soils. Sarkar and Bhattacharyya [53] observed that the mung bean seeds when soaked in suspension of a particular hormones such as cytokinins and auxin brought about reduced root rot incidence in pot experimental set up and also resulted in increased root length, shoot length, dry weight of seedling, and yield as against the control setup. Furthermore, Mohamedy and Baky [54] discovered that biopriming of pea seeds with abscisic acid indicated the highest survival and lowest root rot disease incidence. In addition, highest plant height, enhanced leaves and branches numbers per plant, dry weight of shoots per plant, pod length and diameter, numbers of seed per pod, and lowest chlorosis percentage were observed.

## 5. Role of hormonal priming in resistance against abiotic stress

Different phytohormones have shown effects in improving germination properties and plant growth under abiotic stress (**Table 1**). Abiotic stress is the negative impact of all non-living factors on living organisms in a specific environment [55]. For example, auxin is very important in plant developmental processes such as translocation of carbohydrates as it improves lateral root formation, photosynthetic activities, flowering, and adventitious root development, by so doing, the plant extend its root deep down the soil to obtain water needed for its developmental stages in case of drought stressor [56]. Similarly in case of insufficient nutrient, the adventitious roots can conduct needed nutrients for plant developmental use [57]. According to Roohi and Jameson [58], seed priming with auxin improved the seedling establishment, improved tolerance to drought stress of *Bouteloua gracilis*. This improvement was achieved by enhancing catalase (CAT), superoxide dismutase (SOD), and peroxidase (POD). Fahad et al. [59] observed priming seeds with auxin had improved germination and growth of rice (*Oryza sativa*) and pigeon pea (*C. cajan*), under model of arsenic and cadmium stress.

Seed priming with cytokinin has resulted in the alleviation of abiotic stresses in various plant species such as wheat and soybean by enhancing chlorophyll (Chl) formation thereby improving photosynthetic rate, enhancing membrane stability, and regulate ionic levels under drought stress [26, 60]. However, the further explanation on the mechanisms of how priming with cytokinin mitigate abiotic stress have not been fully understood. However, it may be as a result of its enhancement of chlorophyll formation and enhances stomatal movement thereby improving energy efficiency through photosynthesis [61].

Seed priming with gibberellin in addition with poultry manure has enhanced the growth of pepper (*Capsicum annuum*) plants and improved their salinity tolerance [62]. According to the use of gibberellin to tomato seed (*Solanum lycopersicum*) improved relative leaf water content, stomatal density, and Chl content by mitigating salinity stress.

Wei et al. [63] observed that priming of rice seed with abscisic acid has enhanced the growth rate, survival rate, biomass accumulation, and root formation under of rice under alkaline stress. Similarly, seed priming with abscisic acid improved salinity tolerance thereby leading to enhanced growth properties of rice, wheat, and sorghum [33, 64]. A similar result was reported by Fricke et al. [40] on barley leaves growth through the down regulation of the water loss during transpiration under saline conditions.

Seed priming with salicylic acid have also showed improved growth properties in heavy metals stressed environment Fahad and Bano [39] and Ikhajiagbe and Musa [15]. The application of different levels of salicylic acid was observed to enhance maize (*Zea mays*) yield even under low temperature. Furthermore, garden cress (*Lepidium sativum*) germination and developmental properties as well as seedlings height under salinity stress were enhanced with the application of salicylic acid. Drought stress was also mitigated, while vegetative growth was improved in safflower (*Carthamus tinctorius*) after the application of salicylic acid [65]. Priming of soybean (*Glycine max*) with a combination of ethylene and jasmonic acid had mitigated waterlogging stress by expression of glutathione transferases which led to the promotion of the adventitious root initiation and increasing root surface area [42].

## 6. Role of hormonal priming in resistance against biotic stress

Plants are sessile organisms and therefore cannot move away from its locations in case of environmental stress. For this purpose, plants through evolution have developed series of defense mechanisms. These defense mechanisms can be stimulated either where toxic secondary metabolites are stored; or can be inducible, where defense is activated upon detection of an attack. Plants have the ability to easily detect environmental stress conditions, therefore upon sensing it, they rapidly activate their regulatory or transcriptional machinery, and eventually generate an appropriate response (defense mechanism). Over the years, scientist have gone deep into the research on how plant active their mechanisms against pathogen attack, however, the interplay and impact of different signals to generate defense responses against biotic stress still remain elusive. Seed biopriming with phytohormones has been used in various plant species for the biocontrol of various pathogenic attacks. Abuamsha et al. [66] and Dey et al. [67] applied abscisic acid to the different oil seed rape cultivars which helps in the control of a pathogen causing blackleg disease. The pathogen was observed to be reduced to about 71% after phytohormone application. Muller and Berg [68] reported the role of gibberellin in controlling the damping-off disease in varieties of plant species, especially in cucumber [3].

## 7. Economics of biopriming

Previous researches have shown biopriming with phytohormones to be easier, fast, cheaper, and more environmentally friendly as against the chemical processes. With the enhancement in crop productivity witnessed in biopriming, it have been accepted the potential technique for biocontrol of several plant pathogens. Before now, farmers can only control insect infestation and pathogens attacks through the application of costly and non-ecofriendly pesticides. But with the introduction of hormone priming techniques, plant productivity and pathogens attack can be alleviated through hormone priming. Bio-priming is directly involved in the enrichment of plant development and which improves germination rate, uniformity in plant population, increases water and nutrient use efficiency, eliminates seed borne pathogens, controls pests and diseases. Besides these advantages, bio-priming reduces the hazardous effects on humans caused by the use of fungicides, bactericides, and pesticides by supplementing the chemicals with a sustainable strategy.

## 8. Conclusion and future prospects

Seed priming using plant hormones has shown to be a promising and innovative technique in improving germination parameters as well as growth and yield of varieties of plant species. It has as well showed signs of effectiveness in plant abiotic and biotic stress management. Seed priming with phytohormones result in increased antioxidant secretion and activities, thereby reducing oxidative stress, leading to plant growth and yield enhancement. Therefore, seed hormopriming have the capacity to be utilized for sustainable crop production even under environmental stress. Seed biopriming have also proven to improve seedling health and also improves imbibition rate by breaking dormancy and improving viability. This review shed more light on the successes recorded as a result of using phytohormones in fortifying seeds prior to sowing as it serves as early treatment to plants thereby stimulating all important enzymes at early stage. The information in this review can be used for developing future research on plant growth improvements and would inform modern farmers on the need to consider this important strategy. This emerging strategy has proven to be an effective seed treating technique for many crops. However, phytohormones concentration and priming duration may differs from crop to crop. For example, excessing seed priming and for longer period may lead to desiccation and decomposition of seed or bacteria infestation which makes seeds unviable. Future research at OMICs levels may be required to further explain the mechanisms employed by these phytohormones in seed priming, especially on how it reduces biotic stress in plant. Researches at molecular level is also required to further clarify on pathways involve and influence of priming duration and concentration of the phytohormones.

## Acknowledgements

The researchers are grateful to the Department of Biology and Forensic Science, Admiralty University of Nigeria, Delta State, Nigeria for the encouragement. The mentorship of Prof. Beckley Ikhajiagbe of the Department of Plant Biology and Biotechnology is very much appreciated. I also acknowledge the efforts of Prof. I. J. Dioha, the Dean of Faculty of Science, Admiralty University of Nigeria. Your mentorship is well appreciated.

## Conflict of interest

The authors declare no conflicts of interests.

## Author details

Musa Saheed Ibrahim[1*], Nathan Moses[1] and Beckley Ikhajiagbe[2]

1 Department of Biology and Forensic Science, Admiralty University of Nigeria, Ibusa, Delta State, Nigeria

2 Department of Plant Biology and Biotechnology, University of Benin, Benin City, Edo State, Nigeria

*Address all correspondence to: musa-biology@adun.edu.ng

## References

[1] Kloepper JW, Schroth MN. Plant growth promoting rhizobacteria on radishes. In: Proceedings of the Fourth International Conference on Plant Pathogen Bacteria. Vol. 2. Tours, France: INRA, Gilbert-Clarey; 1978. pp. 879-882

[2] Addicott F, Carns H, Cornforth J, Lyon J, Milborrow BV, Ohkuma K, et al. Abscisic acid: A proposal for the redesignation of abscisin II (dormin). In: Wightman F, Setterfield G, editors. Biochemistry and Physiology of Plant Growth Substances. Ottawa: Runge Press; 1968. pp. 1527-1529

[3] Rhaman MS, Imran S, Rauf F, Khatun M, Carol C, Baskin C, et al. Seed priming with phytohormones: An effective approach for the mitigation of abiotic stress. Plants. 2021;**10**:37. DOI: 10.3390/plants10010037

[4] Sitken A, Yildiz HE, Ercisli S. Effects of plant growth promoting bacteria (PGPB) on yield, growth and nutrient contents of organically grown strawberry. Scientia Horticulturae. 2010;**124**:62-66

[5] Muhei S. Seed priming with phytohormones to improve germination under dormant and abiotic stress conditions. Advances in Crop Science and Technology. 2018;**6**:403-409

[6] Sytar O, Kumar P, Yadav S, Brestic M, Rastogi A. Phytohormone priming: Regulator for heavy metal stress in plants. Journal of Plant Growth Regulation. 2018;**38**:739-752

[7] Hayat R, Ali S, Amara U, Khalid R, Ahmed I. Soil beneficial bacteria and their role in plant growth promotion: A review. Annals of Microbiology. 2010;**60**:579-598

[8] Jeffries P, Gianinazzi S, Perotto S. The contribution of arbuscular mycorrhizal fungi in sustainable maintenance of plant health and soil fertility. Biology and Fertility of Soils. 2003;**37**:1-16

[9] Reddy PP. Recent Advances in Crop Protection. India: Springer; 2013. p. 83

[10] Prabha R, Singh D, Yadav S. Seed biopriming with potential microbial inoculants as sustainable options for stress management in crops. In: Singh D, Prabha R, editors. Microbial Interventions in Agriculture and Environment. Singapore: Springer; 2019

[11] Hasanuzzaman M, Fotopoulos V. Priming and Pretreatment of Seeds and Seedlings: Implication in Plant Stress Tolerance and Enhancing Productivity in Crop Plants. Singapore: Springer; 2019

[12] Masood A, Ellahi N, Batool Z. Causes of low agricultural output and impact on socio-economic status of farmers: A case study of rural Potohar in Pakistan. International Journal of Basic and Applied Science. 2012;**1**:343-349

[13] Taylor A, Allen P, Bennett MA, Bradford JK, Burris JS, Mishra MK. Seed enhancements. Seed Science Research. 1998;**8**:245-256

[14] Di Girolamo G, Barbanti L. Treatment conditions and biochemical processes influencing seed priming effectiveness. Italian Journal of Agronomy. 2021;**7**:8-18

[15] Ikhajiagbe B, Musa SI. The effects of salicylic acid on the germination and early seedling growth of pigeon pea (*Cajanus cajan*). Notulae Scientia Biologicae. 2020;**12**(3):683-692. DOI: 10.15835/nsb12310777

[16] Galhaut L, Lespinay A, Walker DJ, Bernal MP, Correal E, Lutts S. Seed priming of *Trifolium repens* L. improved

germination and early seedling growth on heavy metal-contaminated soil. Water, Air, and Soil Pollution. 2018;**225**:1-15

[17] Patade VY, Khatri D, Manoj K, Kumari M, Ahmed Z. Cold tolerance in thiourea primed capsicum seedlings is associated with transcript regulation of stress responsive genes. Molecular Biology Reports. 2012;**39**:10603-10613

[18] Farooq M, Wahid A, Siddique KHM. Micronutrients application through seed treatments—A review. Journal of Soil Science and Plant Nutrition. 2012;**12**:125-142

[19] Callan NW, Marthre DE, Miller JB. Bio-priming seed treatment for biological control of *Pythium ultimum* pre emergence damping-off in sh-2 sweet corn. Plant Disease. 1990; **74**:368-372

[20] Ibrahim MS, Ikhajiagbe B. The growth response of rice (*Oryza sativa* L. var. FARO 44) in vitro after inoculation with bacterial isolates from a typical ferruginous ultisol. Bulletin of the National Research Centre. 2021;**45**:70. DOI: 10.1186/s42269-021-00528-8

[21] Afzal I, Basra SMA, Ahmad N, Cheema MA, Warriach EA, Khaliq A. Effect of priming and growth regulator treatment on emergence. International Journal of Agriculture and Biology. 2002;**4**:306

[22] Akbari G, Sanavy SA, Yousefzadeh S. Effect of auxin and salt stress (NaCl) on seed germination of wheat cultivars (*Triticum aestivum* L.). Pakistan Journal of Biological Sciences. 2007;**10**:2557-2561

[23] Iqbal M, Ashraf M. Gibberellic acid mediated induction of salt tolerance in wheat plants: Growth, ionic partitioning, photosynthesis, yield and hormonal homeostasis. Environmental and Experimental Botany. 2013;**86**:76-85

[24] Yuan Z, Wang C, Li S, Li X, Tai F. Effects of different plant hormones or PEG seed soaking on maize resistance to drought stress. Canadian Journal of Plant Science. 2014;**94**:1491-1499

[25] Iqbal M, Ashraf M. Seed treatment with auxins modulates growth and ion partitioning in salt stressed wheat plants. Journal of Integrative Plant Biology. 2007;**49**:1045-1057

[26] Mangena P. Effect of hormonal seed priming on germination, growth, yield and biomass accumulation in soybean grown under induced drought stress. Indian Journal of Agricultural Research. 2020;**12**

[27] Angrish A, Kumar B, Datta K. Effect of gibberellic acid and kinetin on nitrogen content and nitrate reductase activity in wheat under saline conditions. Indian Journal of Plant Physiology. 2001;**6**:172-177

[28] Sedghi M, Nemati A, Esmaielpour B. Effect of seed priming on germination and seedling growth of two medicinal plants under salinity. Emirates Journal of Food and Agriculture. 2010;**22**:130-139

[29] Sheykhbaglou R, Rahimzadeh S, Ansari O, Sedghi M. The effect of salicylic acid and gibberellin on seed reserve utilization, germination and enzyme activity of sorghum (*Sorghum bicolor* L) seeds under drought stress. Journal of Stress Physiology & Biochemistry. 2014;**10**:5-13

[30] Watanabe H, Honma K, Adachi Y, Fukuda A. Effects of combinational treatment with ethephon and gibberellic acid on rice seedling growth and carbohydrate mobilization in seeds under flooded conditions. Plant Production Science. 2018;**21**:380-386

[31] Hamza J, Ali M. Effect of seed soaking with GA3 on emergence and seedling growth of corn under salt

stress. Iraqi Journal of Agricultural Science. 2017;**48**:560-566

[32] Zongshuai W, Xiangnan L, Xiancan Z, Shengqun L, Fengbin S, Fulai L, et al. Salt acclimation induced salt tolerance is enhanced by abscisic acid priming in wheat. Plant, Soil and Environment. 2017;**63**:307-314

[33] Gurmani A, Bano A, Khan S, Din J, Zhang J. Alleviation of salt stress by seed treatment with abscisic acid (ABA), 6-benzylaminopurine (BA) and chlormequat chloride (CCC) optimizes ion and organic matter accumulation and increases yield of rice ('*Oryza sativa*' L.). Australian Journal of Crop Science. 2011;**5**:1278-1285

[34] Pouramir-Dashtmian F, Khajeh-Hosseini M, Esfahani M. Improving chilling tolerance of rice seedling by seed priming with salicylic acid. Archives of Agronomy and Soil Science. 2014;**60**:1291-1302

[35] Farooq M, Aziz T, Basra S, Cheema M, Rehman H. Chilling tolerance in hybrid maize induced by seed priming with salicylic acid. Journal of Agronomy and Crop Science. 2008;**194**:161-168

[36] Dai L, Zhu H, Yin K, Du J, Zhang Y. Seed priming mitigates the effects of saline-alkali stress in soybean seedlings. Chilean Journal of Agricultural Research. 2017;**77**:118-125

[37] Onoabhagbe O, Musa SI, Akpeh K, Ekhator PO, Ikhajiagbe B. Germination characteristics of *Sorghum bicolor* (L.) Moench under different pH regimes after chemopriming. Journal of Plant Sciences. 2021;21-33 (in press)

[38] Beckley I, Musa SI, Owalum L, Precious O, Geoffrey O. *In vitro* assessment of elevated soil iron on germinability and germination characteristics of *Sorghum bicolor* (L.) Moench after chemo-priming. Biorxiv. org. 2021. Available from: https://www.biorxiv.org/content/10.1101/2021.11.22.469542v1.full.pdf

[39] Fahad S, Bano A. Effect of salicylic acid on physiological and biochemical characterization of maize grown in saline area. Pakistan Journal of Botany. 2012;**44**:1433-1438

[40] Fricke W, Akhiyarova G, Veselov D, Kudoyarova G. Rapid and tissue-specific changes in ABA and in growth rate in response to salinity in barley leaves. Journal of Experimental Botany. 2004;**5**:1115-1123

[41] Addicott FT, editor. Abscisic Acid. New York: Praeger; 1983

[42] Kim Y, Seo C, Khan A, Mun B, Shahzad R, Ko J, et al. Exo-ethylene application mitigates waterlogging stress in soybean (*Glycine max* L.). BMC Plant Biology. 2018;**18**:254

[43] Montalvo E, Garcia HS, Tovar B, Mata M. Application of exogenous ethylene on postharvest ripening of refrigerated 'Ataulfo' mangoes. LWT - Food Science and Technology. 2007;**40**:1466-1472

[44] Devinar G, Llanes A, Masciarelli O, Luna V. Different relative humidity conditions combined with chloride and sulfate salinity treatments modify abscisic acid and salicylic acid levels in the halophyte *Prosopis strombulifera*. Plant Growth Regulation. 2013; **70**:247-256

[45] Khan W, Prithiviraj B, Smith D. Photosynthetic response of corn and soybean to foliar application of salicylates. Journal of Plant Physiology. 2003;**160**:485-492

[46] Senaratna T, Touchell D, Bumm E, Dixon K. Acetyl salicylic acid (aspirin) and salicylic acid induce multiple stress tolerance in bean and tomato plants. Plant Growth Regulation. 2000;**30**:157-161

[47] Bastam N, Baninasab B, Ghobadi C. Improving salt tolerance by exogenous application of salicylic acid in seedlings of pistachio. Plant Growth Regulation. 2013;**69**:275-284

[48] Jadhav S, Bhamburdekar S. Effect of salicylic acid on germination performance in groundnut. International Journal of Applied Biology and Pharmaceutical Technology. 2011;**3**(4):224-227

[49] Anaya F, Fghire R, Wahbi S, Loutf K. Influence of salicylic acid on seed germination of *Vicia faba* L. under salt stress. Journal of the Saudi Society of Agricultural Sciences. 2015;**17**:1-8

[50] Khan AA. Pre-plant physiological seed conditioning. Horticultural Reviews. 1992;**13**:301-307

[51] Malathi P, Doraisamy S. Effect of seed priming with *Trichoderma* on seed borne infection of *Macrophomina phaseolina* and seed quality in groundnut. Annals of Plant Protection Sciences. 2004;**12**:87-91

[52] Mohamedy EIRSR, Abd Alla MA, Badiaa RI. Soil amendment and seed bio-priming treatments as alternative fungicides for controlling root rot diseases on cowpea plants in nobaria province. Research Journal of Agriculture and Biological Sciences. 2006;**2**:391-398

[53] Sarkar M, Bhattacharyya PK. Biological control of root rot of greengram caused by *M. phaseolina* by antagonistic microorganisms. Journal of Mycopathological Research. 2008; **46**:233-237

[54] Mohamedy EIRSR, Baky AEMMH. Evaluation of different types of seed treatment on control of root rot disease, improvement growth and yield quality of pea plant in nobaria province. Research Journal of Agriculture and Biological Sciences. 2008;**4**:611-622

[55] Sneideris LC, Gavassi MA, Campos ML, Damico-Damiao V, Carvalho RF. Effects of hormonal priming on seed germination of pigeon pea under cadmium stress. Anais da Academia Brasileira de Ciências. 2015;**87**:1847-1852

[56] Ludwig-Muller J. Auxin conjugates: Their role for plant development and in the evolution of land plants. Journal of Experimental Botany. 2011;**62**:1757-1773

[57] Schiefelbein J. Cell-fate specification in the epidermis: A common patterning mechanism in the root and shoot. Current Opinion in Plant Biology. 2003;**6**:74-78

[58] Roohi R, Jameson DA. The effect of hormone, dehulling and seedbed treatments on germination and adventitious root formation in blue grama. Journal of Range Management. 1991;**44**:237-241

[59] Fahad S, Hussain S, Bano A, Saud S, Hassan S, Shan D, et al. Potential role of phytohormones and plant growth-promoting rhizobacteria in abiotic stresses: Consequences for changing environment. Environmental Science and Pollution Research. 2015;**22**: 4907-4921

[60] Iqbal M, Ashraf M, Jamil A. Seed enhancement with cytokinins: Changes in growth and grain yield in salt stressed wheat plants. Plant Growth Regulation. 2006;**50**:29-39

[61] Stoll M, Loveys B, Dry P. Hormonal changes induced by partial root zone drying of irrigated grapevine. Journal of Experimental Botany. 2000;**51**: 1627-1634

[62] Al Taey D. Alleviation of salinity effects by poultry manure and gibberellin application on growth and peroxidase activity in pepper. International Journal of Agriculture Environment and Biotechnology. 2017;**2**:1851-1862

[63] Wei L, Lv B, Wang M, Ma H, Yang H, Liu X, et al. Priming effect of abscisic acid on alkaline stress tolerance in rice (*Oryza sativa* L.) seedlings. Plant Physiology and Biochemistry. 2015;**90**:50-57

[64] Amzallag G, Lerner H. Physiological adaptation of plants to environmental stresses. In: Pessarakli M, editor. Handbook for Plant and Crop Physiology. Boca Raton, NY, USA: Marcel Dekker Inc.; 1990. pp. 557-576

[65] Chavoushi M, Najafi F, Salimi A, Angaji S. Improvement in drought stress tolerance of safflower during vegetative growth by exogenous application of salicylic acid and sodium nitroprusside. Industrial Crops and Products. 2019;**134**:168-176

[66] Abuamsha R, Salman M, Ehlers R. Effect of seed priming with *Serratia plymuthica* and *Pseudomonas chlororaphis* to control *Leptosphaeria maculans* in different oilseed rape cultivars. European Journal of Plant Pathology. 2011;**130**:287-295

[67] Dey R, Pal KK, Bhatt DM, et al. Growth promotion and yield enhancement of peanut (*Arachis hypogaea* L.) by application of plant growth-promoting rhizobacteria. Microbiological Research. 2004;**159**: 371-394

[68] Muller H, Berg G. Impact of formulation procedures on the effect of the biocontrol agent *Serratia plymuthica* HRO-C48 on *Verticillium* wilt in oilseed rape. Biological Control. 2008;**53**: 905-916

Chapter 7

# Phytohormones as Plant Growth Regulators and Safe Protectors against Biotic and Abiotic Stress

*Rizwan Asif, Riffat Yasmin, Madiha Mustafa, Ana Ambreen, Modasrah Mazhar, Abdul Rehman, Shehla Umbreen and Mukhtiar Ahmad*

## Abstract

Plants are playing important role in the planet by providing food for humans and stability in the environment. Phytohormones are key regulators in various physiological processes and among the most important small signaling molecules affecting plant growth and yield production. These biochemical also initiate adaptive responses caused by external stimuli, such as biotic and abiotic stress. Generally, on the basis of physiology, plant hormones roughly fall into two classes. In class one, phytohormones fall which is responsible for plants growth-promoting activities, such as cell division, cell elongation, seed and fruit development, and pattern of differentiation. On the other hand, the second class of hormone play important role in plants' response, such as biotic and abiotic stresses. Some other hormones, such as jasmonates, salicylic acid, brassinosteroids, and strigolactones, also play a key role in plants. Their biochemical signaling network and their crosstalk ability make plant hormones excellent candidates to optimize plant growth and/or mediate abiotic and biotic stresses in agriculture. In the end, the future trends of plant hormone analysis are exploring plant hormones and their applications. We believe the perspective may serve as guidance for the research of plant hormones in the analytical, environmental, and botanical fields.

**Keywords:** plant hormones, growth promoters, stress hormones, biotechnology

## 1. Introduction

Plant hormones or phytohormones are naturally occurring organic substances in miniscule concentrations and exert their action either locally or at distant sites. These chemical messengers with varied chemical properties and specific chemical structures directly influence the growth and development of the whole plant via different biochemical processes. These growth regulators substance coordinate the plant's response simultaneously in abiotic and biotic stresses [1–3]. The occurrence of plant hormones is ubiquitous; they are present in all higher plants and lower plants as well. Their homeostasis in the plant is regulated by synthesis, metabolism, transport to the targeted tissue, and signal transduction which control its activities in the plant. Bioactive hormones are involved in this special type of regulation however intermediate and conjugated forms also play a pivotal role.

The action of plant hormones at the local and distant sites is mediated through different transport mechanisms. Transport of hormones at a distant target is facilitated by loading from the source into the xylem or phloem. In the last decade, several proteins have been identified that act as transporters of hormones at distant sites while short-distance movement of hormones is mediated by symplast, apoplast, or through transcellular mechanism [4]. At one extreme cytokinin get transported from roots to leaves where they prevent senescence and maintain metabolic activity, while at the other extreme the production of the gas ethylene may bring about changes within the same tissue, or within the same cell, where it is synthesized. Chemically plant hormones have a diversified nature comprising of indole, steroids, terpenes, carotenoids, fatty acids, and derivatives of adenine and such diversity reflect their different biological functions [5].

Generally, phytohormones have been divided into two groups on the basis of their functions; group one hormones including, auxin, gibberellin, cytokinin, brassinosteroids, jasmonic acid, and strigolactones. These endogenous signal molecules play a major role in growth-promoting activities by cell division, cell differentiation, elongation, pattern formation, stomatal movement, flowering, and seed germination and development. Hormones in group two are abscisic acid, salicylic acid, and jasmonic acid; mainly involved in biotic and abiotic stress response under different environmental conditions, such as sunlight, soil conditions, soil water, and nutrients [6–8].

Phytohormones do not act alone but in concurrence, or in antagonism, to each other such that the final growth or development represents their net effect. These comprise a unique set of compounds, with distinctive metabolism and properties. Their quality for being the natural compounds with the ability to produce the physiological effect in concentrations lower than those where nutrients and vitamins could not affect these processes makes them unique from the other compounds [9]. Most commonly hormones are classified into the following two categories, as shown in **Figure 1**.

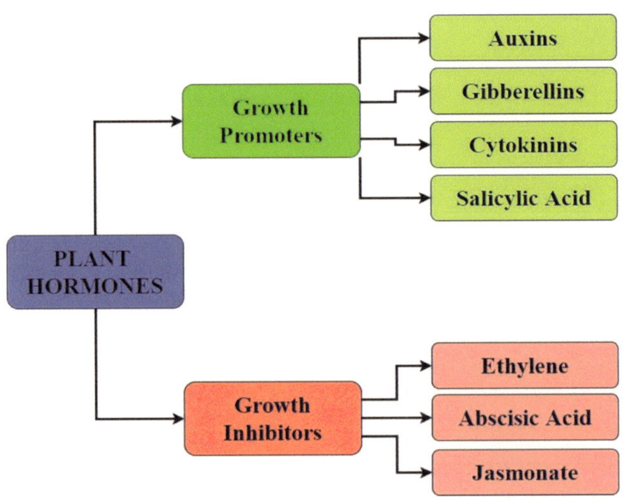


**Figure 1.**
*Classification of phytohormones.*

## 2. Types of hormones

### 2.1 Gibberellin (GA)

Gibberellin (GA) is one of the important plant hormones and it is a tetracyclic di-terpenoid carboxylic acid. It promotes plant growth and development, such as germination of seed, flowering of plants, ripening of fruits, and expansion of leaf while it discourages the growth of trichome. They also play a vital role in the elongation and division of the cells. Its function is also to trigger seed growth and release of seed in dormancy [10]. "To grow or not to grow" is an important verdict for plants to survive. There are requirements of suitable ambiances for plants to grow while in lack of which the growth ceases. So, one of these requirements is the level of growth hormone, GA. That is retained by different synthetic or inactivation of enzymes [11]. GA is also compulsory for the normal growth of roots where its lower concentration is required for maximum root growth other than the shoots. So, GA can be inhibitory for the growth of roots when present in excess quantity [12].

In the 1930s, GA was discovered in Japan while studying a disease related to rice that was with symptoms of excessive growth and yellowing of stems and with lack of production of seeds. GA is with 20 or 19 carbon skeleton ($C_{20}$GA) or ($C_{19}$GA), where GA with 19 carbon skeleton was biologically more active. Presently, about 140 different molecules of GA are known and have been isolated from different microorganisms or plants. GA3, GA4, and GA7 are with the maximum biological characteristics and also, they are commercially available [13]. GA3 is also known as gibberellic acid. The biological active GA is commercially used in agriculture. They are sprayed to increase the size of grapes with no seeds, pears, berries as well as to increase the crop yield under salt stress. It is also practiced in beer brewing to increase the process of malting. In 2016, according to some data the international market scope of GA was estimated at USD 548.9 million [14].

#### *2.1.1 Synthetic pathway of gibberellin*

Gibberellin is chiefly manufactured via the plastidial methylerythritol pathway (MEP) in plants. The biosynthesis of GA in the initial steps is the same in both plants and fungi and is from geranylgeranyl diphosphate (GGPP) to GA-12 aldehyde. Then there is the cyclization of (GGPP) into ent-copalyl diphosphate and to ent-kaurene in plants that is catalyzed by the copalyl diphosphate synthase (CPSp) and the ent-kaurene synthase (KSp) [15]. It is also manufactured by several bacteria and fungi which are associated with various plants in symbiotic or pathogenic relations. In such circumstances, GAs have no evolving utility in the producing organism but perform on the plant host to relief infection by destroying the immunity or on nitrogen-fixing bacteria to adjust the formation of nodules [16].

#### *2.1.2 Gibberellin against heavy metal stress*

Gibberellic acid (GA3) plays an important function for plant growth under salt stress and heavy metal toxicity as well as increases the synthesis of chlorophyll and the action of antioxidant enzymes to prevent lipid peroxidation. GA with calcium (Ca) is added to decrease the salt toxicity like (Ni) toxicity on plant throughput and also to activate different antioxidant enzymes to decrease lipid peroxidation of the cell membrane. They are also involved in regulating the different processes in plants to increase the heavy metal stress [17, 18]. Application of gibberellic acid in *Vicia faba L.*, supported to restore the Cd and Pb-induced reduction in the mitotic index. Due to the application of these growth hormones, GA in plants with heavy metal stress, the ratio of several chromosomal irregularities was expressively reduced. After that, the seeds were harvested from *V. faba* both from heavy metals and gibberellic acid treatment showed a high level of solubilized sugars, proteins, and nucleic acid. Gibberellic acid in lupin plants with Cd stress improved the activity of amylase along with the CAT enzyme. When *Chlorella vulgaris* was exposed to heavy metals, such as Pb and Cd, the application of GA3 amplified the number of cells and level of protein to verify that GA can defend life in water polluted areas [19].

### 2.2 Auxin (IAA)

In the nineteenth century, the idea of a portable substance came into existence that was produced by the leaves and traveled down to encourage the synthesis of the root was capable to control the winding of grass coleoptiles toward the sun. After that these substances were refined, categorized, and were given the name "auxin" from the Greek word "auxin," which means "to grow or to enhance." The naturally existing IAA is indole-3-acetic acid IAA [20]. IAA are organic compounds that are small, low in molecular weight, and constitute the most advanced key and diverse group of phytohormones normally present in all plant types. They are indulged in a number of developing practices by regulating the cell division, such as in the control of shoot building, vascular enlargement, and horizontal root construction [21]. IAA can control senescence, also can react with many pathogens and abiotic or heavy metal stresses. Moreover, it can regulate the production of fruit responses in plants [22].

#### *2.2.1 Synthetic pathway of auxin*

It is essential for the development of the plants that the biosynthesis of auxin be localized in specific tissues but IAA is generally produced in fresh leaves and then it is transported to all through the plant. Many practices have been done to quantify it and its metabolites permitting alteration of the prior opinion of auxin dispersal in a minute quantity of plant tissue having a fresh weight of less than one milligram [23]. IAA biosynthesis in plants is still partly clarified; however, with the use of different isotopes, it is shown that IAA can be synthesized by two important pathways. These pathways are tryptophan dependent and tryptophan independent [21]. In tryptophan dependent pathway, four routes have been identified. In the first route, indolacetamide is converted to IAA by amidohydrolase. In the second route, indole-3-pyruvic acid (IPA) is formed from tryptophan with the help of an aminotransferase which is then converted to indole-3-acetaldehyde by IPA decarboxylase, and then indole-3-acetaldehyde is converted to indole acetic acid by indole-3-acetaldehyde oxidase. These routes are not going parallel often they cross each other. Tryptophan is also converted to tryptamine by trp-decarboxylase and then some proteins help the conversion of tryptamine to indole acetic acid after different steps [24].

In the tryptophan independent pathway, IAA was synthesized in the absence of Trp genes. So, several studies showed that in the absence of tryptophan due to the absence of Trp genes or defective Trp genes, there was evidence of IAA in plants. Studies in Arabidopsis showed that trp3-1 and trp2-1 mutants were defective in tryptophan synthase $\alpha$ and $\beta$ respectively; there was still the production of IAA. It is hypothesized that IAA production might be due to precursors of tryptophan, i.e., indole or indole-3-glycerol phosphate [25].

#### *2.2.2 Auxin against heavy metal stress*

The presence of IAA in plants is also helpful in plant growth in changing environments. In the presence of heavy metal toxicity, it plays a crucial role to tolerate this. In this metal stressed plants, auxin can be provided by the inoculation of microbes that can produce IAA in the rhizosphere of these plants to improve plant production. This metal stress is due to the production of different reactive oxygen species in different locations of the plants, such as root cells, containing peroxisome, mitochondria, plastids, and cytoplasm [26]. In different studies, it is elaborated that the level of IAA is altered endogenously due to the heavy metal stress in roots and shoots. There is shown a positive also a negative relation between heavy metal toxicity and level of IAA. It is also under observation that in response to heavy metal stress, regulation of IAA-producing genes may control the locality and accretion of IAA [27]. Several genes are involved in the relation between IAA and reactive oxygen species to attain homeostasis to adjust the level of $H_2O_2$ by regulating and stimulating the antioxidant enzymes and chlorophyll levels in plants [28]. The abnormal level of metal in the soil causes toxicity in plants and retards the growth and development of the plants due to its accumulation in roots and shoots. The decrease in the rate of growth and development in plants is mainly controlled and maintained by plant hormones like IAA. Indole acetic acid is well known for plant adaptation under heavy metal stress which results in enhanced biomass and production. IAA can be applied to various plants exogenously either directly or through plant growth-promoting rhizobacteria (PGPRs) which produce IAA and in return there is a significant improvement in plant growth under toxic metal concentration [29].

### 2.3 Cytokinins (CKs)

In plant growth and development, the master regulators are known as CKs a phytohormone or plant hormone. Its main function is in the physiology of the cell-like expansion and division of the cell, P and $N_2$ metabolism, maintenance of $H_2O$ balance, the integrity of chloroplast, and senescence. N6-substituted adenine derivatives are known to form by CKs. Seed dormancy can be reduced by using CKs [30]. There are several types of CKs counting—thidiazuron, 6-benzyladenine, kinetin, and 2-isopentenyladenine [31]. CKs are present in abundant types which are different in structure, properties like biochemical and biological activities, and the mode of transportation across plant tissues [32]. CKs are produced in roots and apical meristem and then transported to aerial fragments with the help of absorbent material like minerals via the xylem. In xylem exudate, Zeatin riboside is the utmost copious form of CKs [33]. CKs perform an important role in the growth and adaptation of plants all through the life cycle, such as in the initial phases of reproduction; flowering, seedling, and development [34]. CKs are also used to control the $N_2$ metabolism by increasing the action of nitrate reductase in plants [35].

#### *2.3.1 Synthetic pathway of Cytokinins*

Several studies in plants, such as rice (*Oryza sativa*) and Arabidopsis (*Arabidopsis thaliana*), stated that there are several means to regulate the de novo synthesis of CKs. Synthesis of CKs is due to the nitrogen status signals; one is nitrate-specific signal and the other is a glutamine-related signal. The availability of nitrogen exogenously represents the nitrate-specific signals while assimilated nitrogen status represents the glutamine-related signals. It is known that the CK synthesis is regulated by the nitrogen that is taken up from the soil and is also a key component of nitrogen integration [36]. CKs synthesis have been recognized to be controlled by vital genes. In the initial step that is catalyzed by isopentenyl transferase (IPT), there is a transfer of the isopentenyl group to an adenine nucleotide (ATP, ADP, AMP) from dimethylallyl diphosphate. Then hydroxylation of the methyl group occurs in the isopentenyl side chain by a cytochrome P450. Then in the last step ribose is released and is catalyzed by a phosphoribohydrolase [37].

The other indirect pathway involves the addition of dimethylallyl pyrophosphate (DMAPP) to adenine A37 on tRNA which results in the discharge of CKs nucleotide through degradation of tRNA and by the elimination of the phosphoribosyl by LOG (LONELY GUY). Prenylation of tRNA is catalyzed by tRNA isopentenyltransferase (tRNA-IPT) [38].

### *2.3.2 Cytokinins against heavy metal stress*

When plants are undergone xenobiotic resistance, there is a key role of CKs with saline resistance, drought, light, and temperature signals [39]. Different signaling pathways are there to regulate the concentration of CKs under heavy metal stress to increase plant resistance. *A. thaliana* under arsenic stress is investigated when there are reduced endogenous CKs and the reduced CKs signaling while with the mutant plants exhibited CKs synthesis increased the tolerance of plants against arsenic. It is described that exogenous CKs can promote plant resistance against metal stress [40]. CKs controls morphology, division of cells, and several other substantial routes in the plant. Numerous studies described role of CKs for reduction of heavy metal toxicity in crop plants via biosorption of heavy metals. In higher plants, CKs have been verified to restore the heavy metal-induced decrease in mitotic index resulting in an increase in the number of cells. Furthermore, CKs certainly controlled the photosynthetic mechanism and raised the concentration of various monosaccharides and antioxidants which results in the better existence of plants under heavy metal toxicity [41].

One of the contrivances in a high level of heavy metal is the variation in the level of CKs. It is observed that the concentration of CKs decreases when there is an excess of heavy metal to increase the overall efficiency of plants to cope with this toxicity. Most of the studies agree that the shortage of mineral elements reduces the concentration of CKs in plants [42]. It also improves the tolerance of *C. vulgaris* to Cu, Cd, and Pb due to the activation of the antioxidant defense system, therefore, minimizing the negative values of heavy metal oxidative stress [43]. It is concluded that plant hormones enriched the actions of antioxidant enzymes, i.e., APX, SOD, GR, CAT, and improved the contents of smaller antioxidant molecules, such as glutathione, ascorbate, and proline. CKs protect several proteins and constituents of the process of photosynthesis (carotenes, chlorophylls, and xanthophylls) thus there was a considerable decrease in detrimental effects of heavy metal stress on *A. obliquus* [43].

### *2.3.3 Cross talk among phytohormones*

Phytohormones or plant hormones are small endogenous mediators, such as cytokinin (CK), gibberellin (GA), salicylic acid (SA), brassinosteroid (BR), auxin (IAA), ethylene (ET), jasmonic acid (JA), abscisic acid (ABA), and strigolactone (SL), which coordinate a dual purpose also known as cross talk of plant hormones. Certainly, plant hormones are intermediates that not only direct and organize the progressive practices endogenously but also deliver the environmental incitements to initiate adaptive reactions to biotic & abiotic stress [44].

It was observed that there was an improvement in the differentiation of callus and the number of cells was also increased when CKs were applied with IAA. There was also the contribution of both to uplift the strength of the plant, localization of nutrients, and to increase the grain yield in numerous plants. CKs was found in excessive amount in emerging tissue, such as cambium, root, and shoot tips [45]. Normally, GA and CKs are observed to be antagonists, because both of them opposed the effect of one another on shoot apex, root tips, and elongation of the plant. DELLA proteins are thought to be responsible for such mechanisms where these antagonistic hormones cross talk to function as negative stimulators of GA signals [46]. ABA is known to be a plant stress hormone that accumulates promptly in plants when they are under dehydration or drought stress. A relation exists between ABA and CKs activity for the maintenance of seed development, pre and post-seed development, and stress stimuli. Cross talk concerning ABA and CK is often opposing. When there is an increased concentration of CK, it suppresses the ABA reactions. These antagonistic functional relations are observed for the maintenance of the Arabidopsis under drought conditions although ABA decreases seed sprouting [47]. ET is a gaseous hormone that controls proliferation and expansion of cells, development of fruit, senescence, and various reactions toward biotic and abiotic pressures. The behavior of CKs and ET is generally antagonistic in shoots, where CKs are involved in greening and cell multiplying while ethylene with aging processes, such as maturing, senescence, and the inhibition of cell propagation. Moreover, both of them work cooperatively in different routes, such as maintenance of roots by inhibiting root development. Considerably, CKs positively encourage the production of ET by activating ACC synthase. The production of ET facilitates the CKs to impede hypocotyl elongation in seeds and also to constrain root growth [48]. SA is a hormone of plant origin and performs a vital role against biotrophic pathogens for plant defense in contrast with JA which performs against necrotrophic pathogens. CKs cross talk with SA signals and help in the protected responses via their interactions. CKs increase the SA reactions that help to increase transcription of genes relevant to defense like SID2 and PR1, for the biosynthesis of SA and an indicator or marker gene for SA response respectively [49].

The GA synthesis in an ovule is increased by IAA during fertilization. There is a synergistic effect between auxin and GA signals that regulate fruit development. Their concentrations can be applied externally to regulate the fruit growth and also prompt parthenocarpy [50]. The interpretation of GA and IAA indication cascade has significantly simplified how these hormones manage with each other to control the development of fruit. IAA performs to upstream the level of GA throughout the fruit growth. In Arabidopsis, the fertilization-induced IAA responses or IAA applications activate GA biosynthesis; however, GA applications do not encourage the IAA response [51]. Recently, it is investigated that a high concentration of IAA and the activity of IAA signals enhance ABA-mediated dormancy [52]. Both of them are involved in the maintenance of water status in plants, with opposing performance in the shoots and roots. Indications for water status in the plant have to be integrated to adjust water conductivity in roots and permit modifications in stomatal opening. As a constraint, ABA and IAA signals never exist in a linear fashion but essentially form a system that interconnects through cross talk [53]. IAA and ET cooperatively control numerous developing routes in plants. To date, a whole heap of evidence is existed at the molecular level to promote cross talk between IAA and ET at synthetic, transporter, and signaling levels. That comprises transcriptome profiling datasets to define new entrants for the molecular cross talk. When there is any disturbance in IAA synthesis and transportation, ET helps to promote/deplete the IAA level, or to redistribute it in plants, thereby activating morph-genic reactions. Though, the function of ET in the cross talk is not limited to IAA redistribution [54]. SA and IAA cross talk is fairly obvious from both experimental confirmation and RNA-seq. Data exploration [55]. IAA and SA not only share a common precursor but also play important functions in the maintenance of fruit development and ripening. A metabolic and functional cross talk between them and with other plant hormones occurs in a spatiotemporal fashion to magnificently control the growth of seasonal and non-seasonal fruits [56].

Proofs for protein physiology as a connecting pivot between GA and ABA signaling systems have been progressively evolving both on the functional and on the chromosomal level, mostly in the period of early plant growth. In Arabidopsis, substantial influences of GA and ABA in the signals from hormones have been recognized after the light-reversibility in kernel development. Enhanced production of GA in ABA deficient gene initiated better light-dependent development and concludes an antagonistic association of GA and ABA production in growing and rising seed [57]. ET has a key role in maintaining the developing processes in plants under stress. It is shown to be positively involved with ET when oxygen is deficient. For the elongation of internodes of deepwater rice, GA and ET activity are required. During rice immersing into the water, a decreased level of $O_2$ is recommended to prompt ET synthesis which in turn impede the ABA production. This altered balance between GA and ABA causes elongation of stem induced by GA [58]. GA and SA contribute to the maintenance of many plant reactions. They are concerned to stimulate the expression of proteins involved in pathogenesis. Furthermore, they cooperatively develop plant defenses under biotic and abiotic stresses as SA enhances resistance to abiotic stress in plants. It also has the capability to raise antioxidants and decline the process of lipid peroxidation. In Arabidopsis, seeds were exposed to SA with salt toxicity resulted in the increase of SA because of the activation of two superoxide dismutase, which recovers seed development and upsurges antioxidant capabilities, to enhance the salt tolerance in plants under salt stress [59]. The list of various hormones is given in **Figure 2.**

### 2.4 Abscisic acid (ABA)

Abscisic acid (ABA) is an isoprenoid compound associated with seed dormancy, drought responses, and other growth processes. ABA plays vital roles in plant responses to a range of abiotic stresses such as drought, salinity, high light, nutrient deficiency, and heavy metals. ABA has also been found to be associated with color change in fruits during ripening. ABA at higher concentrations inhibits root growth but in stress conditions, it also plays a vital role in the elongation of the root. Various environmental factors regulate the levels of ABA, including seed maturation, the genotype of plant, water and soil conditions. ABA concentrations are generally increased in nutrient deficiency and decreased at higher temperatures 40°C. Root tissue generally contains lower concentrations of ABA than leaves, dehydration of detached roots from various species, ages, and branching orders also stimulate ABA synthesis [60].

The biosynthesis, catabolism, transport, downstream response, and modulation of ABA have been extensively investigated in angiosperms. ABA is primarily synthesized from carotenoids under the catalytic action of various enzymes such as b-carotene

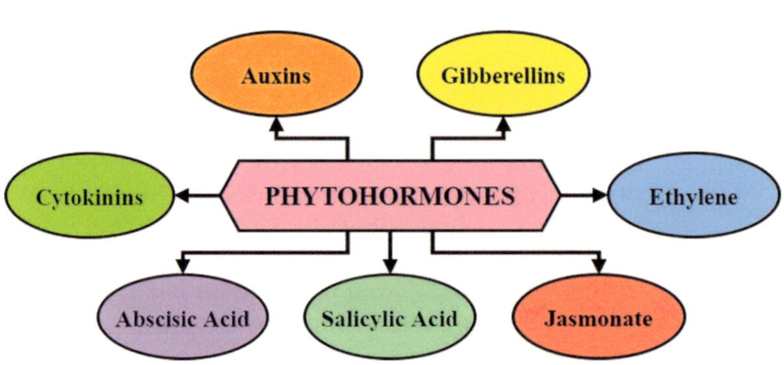


**Figure 2.**
*List of various kinds of phytohormones.*

hydroxylases, zeaxanthin epoxidase (ZEP, ABA1), 9-cis-epoxycarotenoid dioxygenase (NCEDs), short-chain alcohol dehydrogenase/reductases (SDRs, such as ABA2), abscisic aldehyde oxidases (AAOs), molybdenum cofactor sulfurase (MOCO, ABA3), and ABA4. ABA levels are regulated by two major pathways—hydroxylation and esterification mediated by four CYP707As and eight glucosyltransferases (UGTs). The inactivated ABA-glucosyl ester (ABA-GE) conjugation is a storage form of ABA and the site can be cleaved by b-glucosidases (BGLUs) [61].

#### *2.4.1 Abscisic acid against heavy metal stress*

ABA is one of the foremost phytohormones driving plant resistance to toxic metals and metalloids, such as Cd, and Pb. Mechanisms of ABA in response to heavy metals and metalloids stresses in non-angiosperm plant lineages is still limited and not completely understood [62], however, ABA act in different ways in response to heavy metal stress, including by alleviating toxic metal and metalloid stress via ABCGs, PSE1, and WRKY13, limiting their uptake, altering the distribution between roots and shoot and promoting chelation and vacuolar sequestration [7–9].

### 2.5 Jasmonic acid (JA)

Jasmonic acid is a signaling chemical that mediates the number of biotic and abiotic stress process in plants, such as fruit ripening and seed germination, wounding, and ultraviolet radiation. It is produced from linolenic acid but activated after conjugation with isoleucine which permits it to join with COI1and act as a JA receptor. The first role of JA as a senescence-promoting was observed from a compound isolated from wormwood cause rapid loss of chlorophyll in oat [63]. JA also stimulates the secretion of volatile oil in plants which have antimicrobial properties. In a current study remarkable role of JA acid was observed in the regulation of the life cycle in plants [64]. JA is a naturally grown regulator and is extensively found in plants. Normally, JA does not work in isolated form, and extensively cross talk behavior was studied with other hormones. JA's role in a hot climate as a water conserver was also observed through stomatal closure. In addition, JA also helps to cope with drought stress by promoting some enzymes, such as superoxide dismutase (SOD), peroxidase (POD), catalase (CAT), proline. The role of JA against fungal pathogens and some other plants pathogen was also observed [65]. JA not only works as a plant's growth regulator but also stimulates the immune system in plants. The list of hormones with their origin and functions is mentioned in **Table 1**.

### 2.6 Salicyclic acid (SA)

Salicyclic acid is one of the most important plant hormones naturally produced in plants within the cytoplasm of the cell. SA is a key player in the regulation of very important functions, such as photosynthesis, growth, and even in defense of plants [67]. Different past studies demonstrated the role of SA in plants against biotic and abiotic stress. SA stimulates the SAR mechanism in plants which activates pathogenesis-related proteins which work against different kinds of phytopathogens, such as fungus and bacteria. It also induces a variety of metabolic processes in plants and also regulates plant and water relations [68]. The role of SA was also observed in different kinds of signaling which leads to gene expression and protein synthesis. Similarly, SA also works with other hormones in cross talk. The SA treatment in cucumber or tobacco plants induced heavy metal-like cupper tolerance ability. It also induced cd tolerance ability in plants but the exact mechanism is still unknown [66]. The hormones of various names with their functions are given in **Figure 3**.

| Phytohormones | Origin/precursor of Synthesis | Functions | | | | | | | References |
|---|---|---|---|---|---|---|---|---|---|
| | | Germination | Growth | Flowering | Fruit development | Abscission | Seed dormancy | Defense mechanism | |
| Gibberellin | Young shoots, growing seeds/ Glyceraldehyde 3-phosphate | Yes | Yes | Yes | Yes | No | Yes | No | [10] |
| Auxin | Young leaves, developing Fruit/ Tryptophan or IAA | No | Yes | Yes | Yes | No | No | No | [22] |
| Cytokinins | Root tip, young leaves, growing seeds/Adenine | No | Yes | Yes | Yes | No | No | Yes | [35] |
| Abscisic acid | Roots & Leaves/Carotenoids | No | No | No | No | Yes | Yes | No | [60] |
| Jasmonic acid | Various tissues/Polyunsaturated Fats/ Linolenic acid | Yes | No | No | Yes | No | No | Yes | [65] |
| Salicyclic acid | Various tissues/Phenylalanine | No | Yes | No | No | No | No | Yes | [66] |

**Table 1.**
*List of phytohormones with their origin and functions.*

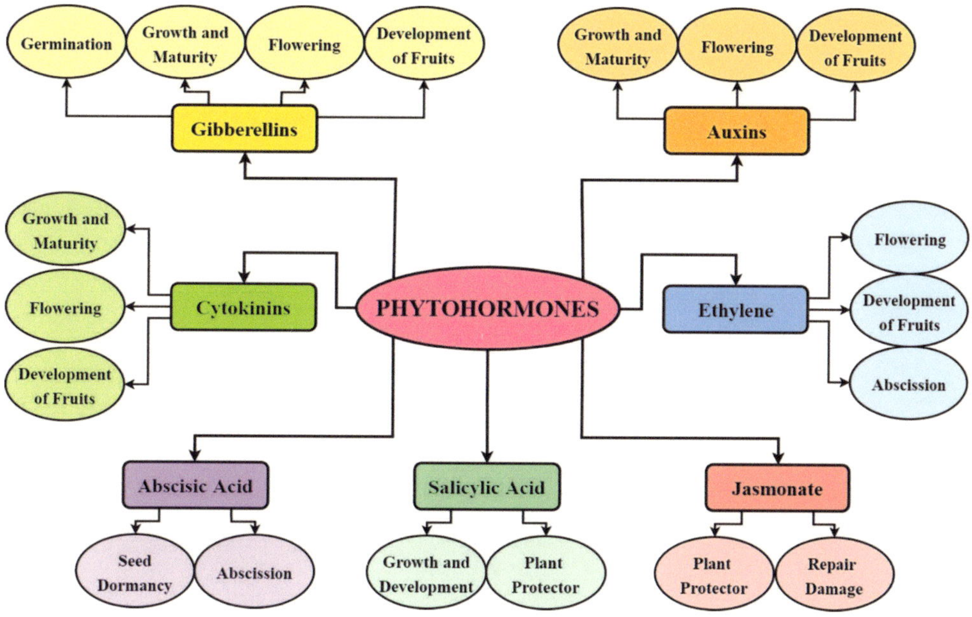


**Figure 3.**
*Phytohormones and their role in plants.*

## 3. Conclusion

Growth is an essential property for every living organism and is usually regulated by various external and internal factors. Generally, plants attain this property via synthesizing a small amount of chemical substance known as phytohormones. These chemical substances trigger biochemical changes that ultimately initiate several growth changes in plants, such as the formation of flowers, roots, stems, and fruits. As a result, these processes increase the yield. Some phytohormones also play important role in a plant's life from dormancy to senescence. Consequently, perform a vital role in agriculture and horticulture, etc. Conclusively, phytohormone regulates the physiology of plants but the information about the molecular mechanisms still remains unclear.

## Author details

Rizwan Asif[1*], Riffat Yasmin[2], Madiha Mustafa[1], Ana Ambreen[1], Modasrah Mazhar[1], Abdul Rehman[1], Shehla Umbreen[1] and Mukhtiar Ahmad[3]

1 Department of Eastern Medicine, Qarshi University, Lahore, Pakistan

2 Riphah College of Rehabilitation and Allied Health Sciences (RCRAHS), Riphah International University, Faisalabad, Pakistan

3 University College of Conventional Medicine, The Islamia University of Bahawalpur, Bahawalpur, Pakistan

*Address all correspondence to: asifrizwan429@yahoo.com

## References

[1] Asami T, Nakagawa Y. Preface to the special issue: Brief review of plant hormones and their utilization in agriculture. Journal of pesticide science. 2018;**43**(3):154-158

[2] Shah SH et al. Role of exogenously applied plant growth regulators in growth and developmen of edible oilseed crops under variable environmental conditions: A review. Journal of Soil Science and Plant Nutrition. 2021;**21**:3284-3308

[3] Wang L et al. Recent developments and emerging trends of mass spectrometric methods in planthormone analysis: A review. Plant Methods. 2020;**16**(1):1-17

[4] Abualia RE, Benkova LB. Transporters and mechanisms of hormone transport in Arabidopsis. Advances in Botanical Research. 2018;**87**:115-138

[5] Srivastava LM. Plant Growth and Development: Hormones and Environment. Amsterdam, Netherlands: Elsevier; 2002

[6] Müller M, Munné-Bosch S. Hormonalimpact on photosynthesis and photoprotection in plants. Plant Physiology. 2021;**185**(4):1500-1522

[7] Jiang K, Asami T. Chemical regulators of plant hormones and their applications in basic research and agriculture. Bioscience, Biotechnology, and Biochemistry. 2018;**82**(8): 1265-1300

[8] Bari R, Jones JD. Role of plant hormones inplant defence responses. Plant Molecular Biology. 2009;**69**(4): 473-488

[9] Davies PJ. Plant Hormones: Biosynthesis, Signal Transduction, Action!. Netherlands: Berlin: Springer Science & Business Media; 2004

[10] Emamverdian AY, Ding MF. The role of salicylic acid and gibberellin signalingin plant response to abiotic stress with an emphasis on heavy metals. Plant Signaling & Behavior. 2020;**15**(7):1777372

[11] Takehara S et al. A common allosteric mechanism regulates homeostatic inactivation of auxin and gibberellin. Nature Communications. 2020;**11**(1):1-10

[12] Barker R et al. Mapping sites of gibberellin biosynthesis in the Arabidopsis root tip. New Phytologist. 2021;**229**(3):1521-1534

[13] Camara MC et al. Current advances ingibberellic acid (GA 3) production, patentedtechnologiesand potential applications. Planta. 2018;**248**(5): 1049-1062

[14] Kildegaard KR et al. Tailored biosynthesis of gibberellin plant hormones in yeast. Metabolic Engineering. 2021;**66**:1-11

[15] Tudzynski B et al. The P450-4 gene of Gibberella fujikuroi encodes ent-kaurene oxidase in thegibberellin biosynthesis pathway. Applied and Environmental Microbiology. 2001;**67**(8):3514-3522

[16] Hedden P. The current status of research on gibberellin biosynthesis. Plant and Cell Physiology. 2020;**61**(11):1832-1849

[17] Siddiqui MH, Al-Whaibi MH, Basalah MO. Interactive effect of calcium and gibberellin on nickel tolerance in relation to antioxidant systems in Triticum aestivum L. Protoplasma. 2011;**248**(3):503-511

[18] Colebrook EH et al. The role of gibberellin signalling in plant responses to abiotic stress. Journal of Experimental Biology. 2014;**217**(1):67-75

[19] Saini SN, Kaur PPK. Phytohormones: Key players in the modulation of heavy metalstress tolerance in plants. Ecotoxicology and Environmental Safety. 2021;**223**:112578

[20] Tromas A, Perrot-Rechenmann C. Recent progress in auxin biology. Comptes Rendus Biologies. 2010;**333**(4): 297-306

[21] Woodward AW, Bartel B. Auxin: Regulation, action, and interaction. Annals of Botany. 2005;**95**(5):707-735

[22] Wang S, Fu J. Insights into auxin signaling in plant–pathogen interactions. Frontiers in Plant Science. 2011;**2**:74

[23] Ljung K, Bhalerao RP, Sandberg G. Sites and homeostatic control of auxin biosynthesis in Arabidopsis during vegetative growth. The Plant Journal. 2001;**28**(4):465-474

[24] Sugawara S et al. Biochemical analyses of indole-3-acetaldoxime-dependent auxinbiosynthesis in Arabidopsis. Proceedings of the National Academy of Sciences. 2009;**106**(13):5430-5435

[25] Ouyang J, Shao X, Li J. Indole-3-glycerol phosphate, a branchpoint of indole-3-acetic acid biosynthesis from the tryptophan biosynthetic pathway in *Arabidopsis thaliana*. The Plant Journal. 2000;**24**(3):327-334

[26] Singh H et al. Auxin metabolic network regulates the plant response to metalloids stress. Journal of hazardous materials. 2021;**405**:124250

[27] Bruno L et al. In Arabidopsis thaliana cadmium impact on the growth of primary root by altering SCR expression and auxin-cytokinin cross-talk. Frontiers in Plant Science. 2017;**8**:1323

[28] Fahad S et al. Phytohormones and plant responses to salinity stress: A review. Plant Growth Regulation. 2015;**75**(2):391-404

[29] Zhang C et al. Exogenous indole acetic acid alleviates Cd toxicity in tea (Camellia sinensis). Ecotoxicology and Environmental Safety. 2020;**190**:110090

[30] Emamverdian A et al. Mechanisms of selected plant hormones under heavy metal stress. Polish Journal of Environmental Studies. 2021;**30**(1): 497-508

[31] Bekircan T et al. Effect of cytokinins on in vitro multiplication, volatiles composition and rosmarinic acid content of Thymus leucotrichus Hal. Shoots. 3 Biotech. 2018;**8**(3):1-9

[32] Jameson PE, Song J. Cytokinin: A key driver of seed yield. Journal of Experimental Botany. 2016;**67**(3): 593-606

[33] Mangieri MA et al. Cytokinins: A key player in determining differences in patterns of canopy senescence in stay green and fast dry-down sunflower (Helianthus annuus L.) hybrids. European Journal of Agronomy. 2017;**86**:60-70

[34] Nguyen HN et al. Cytokinin activity during early kernel development corresponds positively with yield potential and later stage ABA accumulation in field-grown wheat (Triticum aestivum L.). Planta. 2020;**252**(5):1-16

[35] Sýkorová B et al. Senescence-induced ectopic expression of the A. tumefaciens ipt gene in wheat delays leaf senescence, increases cytokinin content, nitrate influx, and nitrate reductase activity, but does not affect grain yield. Journal of Experimental Botany. 2008;**59**(2):377-387

[36] Sakakibara H. Cytokinin biosynthesis and transport for systemic nitrogen signaling. The Plant Journal. 2021;**105**(2):421-430

[37] Zhao Y et al. A wheat allene oxide cyclase gene enhances salinity tolerance via jasmonate signaling. Plant Physiology. 2014;**164**(2):1068-1076

[38] Hrtyan M et al. RNA processing in auxinand cytokinin pathways. Journal of Experimental Botany. 2015;**66**(16): 4897-4912

[39] Pavlů J et al. Cytokinin at the crossroads of abiotic stress signalling pathways. International Journal of Molecular Sciences. 2018;**19**(8):2450

[40] Mohan TC et al. Cytokinin determines thiol mediated arsenic tolerance and accumulation. Plant Physiology. 2016;**171**(2):1418-1426

[41] Zhou MX et al. Effect of NaCl on proline and glycinebetaine metabolism in Kosteletzkya pentacarpos exposed to Cd and Zn toxicities. Plant and Soil. 2019;**441**(1):525-542

[42] Veselov D et al. Role of cytokinins in stress resistance of plants. Russian Journal of Plant Physiology. 2017;**64**(1): 15-27

[43] Piotrowska-Niczyporuk A et al. Exogenously applied auxins and cytokinins ameliorate lead toxicity by inducing antioxidant defence system in green alga *Acutodesmus obliquus*. Plant Physiologyand. Biochemistry. 2018;**132**:535-546

[44] Vanstraelen M, Benková E. Hormonal interactions in the regulation of plant development. Annual Review of Cell and Developmental Biology. 2012;**28**:463-487

[45] Checker VG et al. Role of phytohormones in plant defense: Signaling and cross talk. In: Molecular Aspects of Plant-Pathoge Interaction. Berlin: Springer; 2018. pp. 159-184

[46] Davière JM, Achard P. A pivotal role of DELLAs in regulating multiple hormone signals. Molecular Plant. 2016;**9**(1):10-20

[47] Huang X et al. The antagonistic action of abscisic acid and cytokinin signaling mediates drought stress response in Arabidopsis. Molecular Plant. 2018;**11**(7):970-982

[48] Zdarska M et al. ETR1 integrates response to ethylene and cytokinins into a single multistep phosphorelay pathway to control root growth. Molecular Plant. 2019;**12**(10):1338-1352

[49] Naseem M et al. Integrated systems view on networking by hormones in Arabidopsis immunity reveals multiple crosstalk for cytokinin. The Plant Cell. 2012;**24**(5):1793-1814

[50] Matsuo S et al. Roles and regulation of cytokinins in tomato fruit development. Journal of Experimental Botany. 2012;**63**(15):5569-5579

[51] Hu J et al. The interaction between DELLA and ARF/IAA mediates crosstalk between gibberellin and auxin signaling to control fruit initiation in tomato. The Plant Cell. 2018;**30**(8): 1710-1728

[52] Ye J et al. Knockdown of SlNL33 accumulates ascorbate, enhances disease and oxidative stress tolerance in tomato (Solanum lycopersicum). Plant Growth Regulation. 2019;**89**(1):49-58

[53] Behnke K et al. RNAi-mediated suppression of isoprene emission in poplar transiently impacts phenolic metabolism under high temperature and high light intensities: A transcriptomic and metabolomic analysis. Plant Molecular Biology. 2010;**74**(1-2):61-75

[54] Zemlyanskaya EV et al. Deciphering auxin ethylene crosstalk at a systems level. International Journal of Molecular Sciences. 2018;**19**(12):4060

[55] Tiwari J et al. Glycerol micellar catalysis:An efficient

multicomponent-tandem green synthetic approach to biologically important 2,4-disubstituted thiazole derivatives. ChemistrySelect. 2018;**3**(41):11634-11642

[56] Pérez-Llorca M et al. Biosynthesis, metabolism and function of auxin, salicylic acid and melatonin in climacteric and nonclimacteric fruits. Frontiers in Plant Science. 2019;**10**:136

[57] Golldack D et al. Gibberellins and abscisic acid signal crosstalk: Living and developing under unfavorable conditions. Plant Cell Reports. 2013;**32**(7):1007-1016

[58] Bashar KK et al. Phytohormone-mediated stomatal response, escape and quiescenc strategies inplants under flooding stress. Agronomy. 2019;**9**(2):43

[59] Fayez KA, Bazaid SA. Improving drough and salinity tolerance in barley by application of salicylic acid and potassium nitrate. Journal ofthe Saudi Society of Agricultural Sciences. 2014;**13**(1):45-55

[60] Davies PJ. The plant hormones: Their nature, occurrence, and functions. In: Plant Hormones. Berlin: Springer; 2010. p. 15

[61] Finkelstein R. Abscisic acid synthesis and response. The Arabidopsis book/American Society of Plant Biologists. 2013;**11**:e0166

[62] Shi WG et al. Abscisic acid enhances lead translocation from the roots to the leaves and alleviates its toxicity in Populus× canescens. Journal of Hazardous Materials. 2019;**362**:275-285

[63] Kelley DR, Estelle M. Ubiquitin-mediated control of plant hormone signaling. Plant Physiology. 2012;**160**(1): 47-55

[64] Gomi K. Jasmonic Acid: An Essential Plant Hormone. Basel, Switzerland: Multidisciplinary Digital Publishing Institute; 2020

[65] Yang J et al. The crosstalks between jasmonic acid and other plant hormone signaling highlight the involvement of jasmonic acid as a core component in plant response to biotic and abiotic stresses. Frontiers in Plant Science. 2019;**10**:1349

[66] Janda T, Szalai G, Pál M. Salicylic acid signalling in plants. Basel, Switzerland: Multidisciplinary Digital Publishing Institute; 2020

[67] Janda M, Ruelland E. Magical mystery tour: Salicylic acid signalling. Environmental and Experimental Botany. 2015;**114**:117-128

[68] Mohamed HI, El-Shazly HH, Badr A. Role of salicylic acid in biotic and abiotic stresstolerance in plants. In: Plant Phenolics in Sustainable Agriculture. Berlin: Springer; 2020. pp. 533-554

Chapter 8

# Role of Endogenous and Exogenous Hormones in Bioactive Compounds Production in Medicinal Plants via In Vitro Culture Technique

*Majid Ibrahim*

## Abstract

The natural compounds produced in plants are classified into two major groups (Primary and secondary metabolic compounds). These compounds are the precursor materials for thecompounds of the second group, which are represented by secondary metabolites, most of which produce from three main compounds: shikimic acid, acetate, and fatty acids. Primary metabolites are the basic units in the metabolism of secondary compounds. Tissue cultures of plants are used to produce large quantities of secondary metabolic products, although cultures of callus and cell suspensions often do not produce higher levels of the whole plant. Therefore, some technologies were used to increase the production of secondary metabolites by plant tissue culture techniques through the selection of high-production cells. The growth of plant cells in tissue cultures occurs when the requirements for division and growth are available for them from nutrients, growth regulators, and any other additives that all affect the metabolic activities within the cells. To achieve optimal productivity of secondary metabolites, it is preferable to produce cells in a medium that is optimal for increasing biomass. Plant growth regulators such as auxins and cytokinins affect cell division, various metabolic processes, and plant growth in tissue cultures.

**Keywords:** auxin, cytokinin, *in vitro* culture, plant hormones, secondary metabolites

## 1. Introduction

The plants can produce certain bioactive compounds that are mostly affected by the chemical and physical environments in which they develop. Some searches announced that plant growth regulators and light are important factors stimulating the growth, development (organogenesis), and production of plant compounds, including both primary and secondary products. In addition, plant growth regulators were applied for callus induction, and adjusting the metabolite content, carbon sources, suspension culture, temperature, pH, medium type and ammonium nitrate ($NH_4NO_3$) concentrations plays an important role in the formation of plant primary and secondary products [1–3]. This chapter elicits the role of endogenous and exogenous hormones are active for enhancing the following stages–cellular division stage, cell enlargement stage, exponential stage, steady stage, and reduce biomass as well as secondary products content in medicinal plant.

## 2. Secondary products

The natural compounds produced in plants are classified into two major groups. The first group includes compounds that enter into primary reactions or primary metabolic compounds. This name refers mostly to the metabolic processes that produce simple basic carboxylic acids, amino acids, sugars, lipids, proteins, and nucleic acids. These compounds are the precursor materials for the compounds of the second group, which are represented by secondary metabolites, most of which produce from three main compounds: shikimic acid, acetate, and fatty acids. Primary metabolites are the basic units in the metabolism of secondary compounds, which are divided into several different groups and generally include terpenes, phenols, alkaloids, glycosides, tannins, resins, and others [3, 4].

### 2.1 Glycosides

These compounds are bioactive and important substances in the defense and metabolic system of medicinal plants. These compounds are helped by these plants to complete their life cycle by protecting them from biotic stresses (defense against infection with bacteria, viruses, fungi, nematodes, rodents, etc.). As well as its important role in the treatment of many diseases that affect humans and animals [3, 5, 6]. Glycosides consist of two molecules, one of which is a sugar called the glycon, which is monosaccharides, disaccharides, or polysaccharides. This part of the sugar works to transport the glycoside molecule across the cell membranes, so it has the properties of pharmacokinetics. As for the other part, it is called an aglycon, which may be an alcohol, an aldehyde, a ketone, or an ester, and it is attributed to this part of the physiochemical effectiveness (**Figure 1**). The glycon is attached to the aglycon part by several chemical bonds, which may be an oxygen, sulfur, or carbon bond. Glycoside compounds include steroids, anthraquinones, tannins, and saponins [4–6].

### 2.2 Alkaloids

Alkaloids are a group of low-molecular-weight basic organic compounds, whose molecule contains one or more nitrogen atoms linked to heterogeneous rings, so the alkaloids do not share a specific chemical composition. The human knew plants containing alkaloid compounds 3000 years ago and used their extracts to heal from

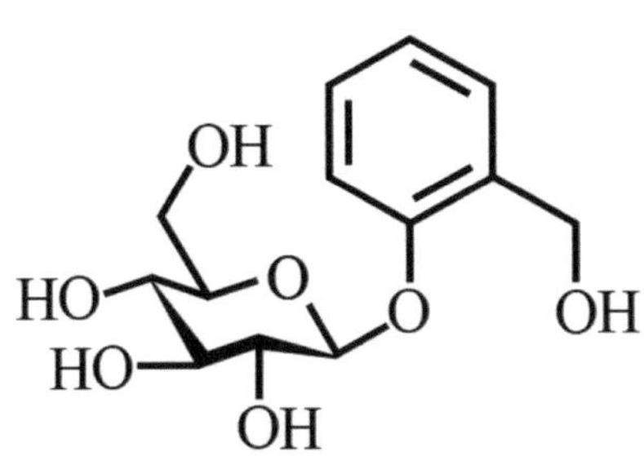


**Figure 1.**
*Glycosides compound.*

diseases, treat wounds, or make poisons used in hunting, defending him, or religious rituals. The first process of isolation of alkaloid compounds was the isolation of alkaloid morphine from the papaver plant by the German scientist Derosnein in 1803 AD. Then it was followed by the isolation of many alkaloid compounds that saved the lives of millions of people from incurable diseases or those that contributed to alleviating the pain of surgical operations [4, 6]. Alkaloids are usually found free or in the form of salts of some organic acids such as citric acid, tannic acid, and tartaric acid. Alkaloids are produced by bacteria, fungi, and higher plants and are found in all parts of the plant, such as hyoscine alkaloid in tobacco, in seeds, such as strychnine alkaloid in emetic walnut, in the roots, such as glycyrrhizin alkaloid in licorice, in the bark, such as cinchonine alkaloid in cinchona, in the fruits are like capsaicin alkaloid in the black pepper plant, or in Latex like the papaverine alkaloid in the poppy plant. Alkaloids are divided into several groups according to the chemical structure of the basic ring in the alkaloid molecule into the group of amine alkaloids, pyridine and piperidine, tropane alkaloids, quinoline alkaloids, purine alkaloids, isoquinoline alkaloids, indole alkaloids, phenolic alkaloids, tropolone alkaloids and tropolone alkaloids (**Figure 2**) [3, 4].

### 2.3 Phenols

Phenolic compounds are the second largest group of secondary metabolites in plants after the alkaloid group. A simple phenolic molecule contains a benzene ring to which one or more hydroxyl groups are attached. These compounds are found in both higher and lower plants (such as ferns, mosses, and many microorganisms). Phenols are also called aromatic compounds because of their distinctive smell, and they are sometimes called closed ring compounds because they contain a benzene ring (**Figure 3**). These compounds are characterized by the presence of a hydroxyl group (OH) directly attached to the aromatic ring. Sometimes several different

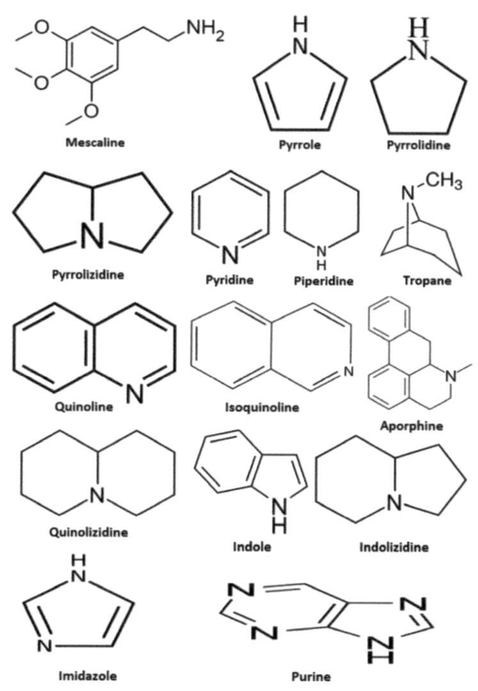


**Figure 2.**
*Alkaloid compounds.*

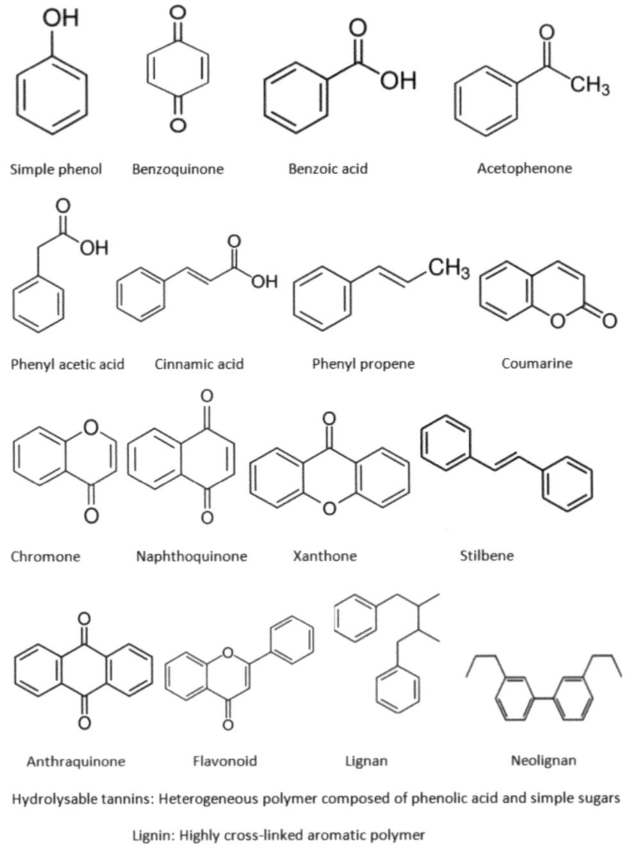


**Figure 3.**
*Phenolic compounds.*

groups are attached to the phenolic compound, such as the hydroxyl group OH, the carboxyl COOH, and the methyl $CH_3$ [3, 6]. The phenolic compounds may exist in the form of an open chain or aliphatic (noncyclic) compounds. Most of the phenolic compounds are not found free inside plant cells but are bound with one or several molecules of sugars to be in the form of glycosidic compounds. There are also some phenolic compounds linked with lipopolysaccharides by a glycoester with one of the OH or COOH groups to form glycolipids that are stored in the cell vacuoles. Some amino acids, such as Tryptophan, Tyrosine, and Phenylalanine, are classified as closed ring organic phenolic compounds, due to the similarity of the method of metabolism of these acids with phenolic compounds [4, 6].

### 2.3.1 Simple phenols

Phenols are the basic material in the biosynthesis of lignin. Phenols also play an important role in regulating plant growth and development by affecting the effectiveness of hormones and their control over the effectiveness of the formation of some enzymes. They represent one of the forms of energy compounds stored by the plant and nutrients that can be utilized when needed. It acts as an antioxidant that hinders the oxidation of chlorophyll and hormones and stabilization in the stabilization of some vital compounds. It also participates in the oxidation and respiration processes. The most important groups of phenolic compounds in higher plants are groups of cinnamic acid, coumarin, lignin, phenolic carboxylic acids, and flavonoids derivatives [4, 7].

#### *2.3.2 Polyphenols (flavonoids)*

Polyphenols or flavonoids are heterocyclic oxygen compounds of essential importance in plant life and are sometimes called Anthoxanthins. Flavonoids are distinguished by their crystalline form and yellow color, which derives from the Latin word flavus. These compounds are found in higher plants, especially some families such as Compositae, Cucurbitaceae, and Umbelliferae, in plant parts such as roots, leaves, flowers, and fruits. Flavonoids are used as an antiviral, general anti-inflammatory, anti-bacterial, and increase the level of immunomodulation. They act as antioxidants, relieve pain, swelling and bruising, stimulate blood circulation, and reduce cholesterol levels in the blood. Flavonoids are divided into several groups: Flavone, Flavonol, Flavonone, Isoflavone, Chalcone, Aurone, Anthocyanin, and Betacyanins [3, 4, 7].

### 2.4 Oils

#### *2.4.1 Volatile oils*

Volatile oils are organic compounds characterized by their volatility or evaporation without decomposing when exposed to heating or at room temperature. They are also called ethereal oils because of their solubility in alcohols, especially ether. Essential oils are so named for their pleasant aromas, and essential oils are so named because they are included in the basic human diet, and the cellular enzyme system in the human body cannot produce it. These oils are spread in more than 2000 plants represented by sixty families, the most important of which are Lauraceae, Labiatae, Umbelliferae, Rutaceae, Compositae, Myrtaceae, Pinaceae, and Oleaceae [3, 8]. The volatile oils in the plant act as pleasant aromas to attract insects to complete the pollination process and at the same time work to exclude other insects. Some of these oils are poisonous and others have a pungent taste that is unpalatable to insects and rodents, meaning that these oils act as an open immune system to defend the plant itself. Volatile oils have an important role in allelochemicals to reduce competition from other plants for light, water, and soil nutrients. These oils are used in the treatment of many diseases, such as eucalyptus oil, which treats inhalation shortness of breath, and bronchitis, which is Antispasmodic. Peppermint oil is used as a mouthwash and antiseptic gargle, Thymol oil is used to treat skin problems, and clove and thyme oil are used as antiseptics because they contain a high percentage of phenolic compounds. Dill oil is used as a carminative, castor oil is used as a laxative, and watercress oil is used as a cholesterol reducer. The volatile oils are divided into groups: Alcoholic, Aldehyde, Ketone, Phenolic, Nitrogenic, and Sulfuric oils [4, 8].

#### *2.4.2 Fixed oils*

Vegetable fixed oils are less dense liquids than water and do not mix with water. It is biologically built in the places of its production and is not transmitted from one plant member to another. It is often produced and stored in seeds and fruits, and a small percentage is produced in the bark and leaves. The most important fixed vegetable oils are corn, sunflower, safflower, cottonseed, sesame, soybean, linseed, archies, and olive oil. There are other fixed oils consisting of esters of unsaturated fatty acids such as linoleic acid, linolenic acid, arachidonic acid with triglyceride alcohol, and others [4, 9]. Fixed oils are distinguished by their high nutritional value, which are used in weight gain programs and in the manufacture of food and energy drinks, among others. The increasing demand for its use in every home is due to its lack of contribution to raising the level of cholesterol and triglycerides in the blood that causes atherosclerosis and heart disease. Fixed oils are also used in the treatment of some diseases, especially spasms, muscle pain, and rheumatism, or as sterilizers or moisturizers for skin cracks resulting from infection with some fungi or bacteria, burns, or sunburn. It is also used as carriers or organic solvents for active compounds in the manufacture of some creams and ointments. These oils are also used in the manufacture of washing and cosmetics, among others. These oils are characterized by being odorless, tasteless, slightly yellowish, hydrophobic, and non-polar compounds that do not dissolve in water but dissolve in organic solvents such as chloroform, benzene, and ether. Fixed oils are divided according to the fatty acids that bind to glycerol into monoglyceride, diglyceride, and triglyceride groups [3, 4].

### 2.5 Resins

Resins are solid or semi-solid organic compounds of different and chemically complex compositions that result from the oxidation of volatile oils. Resins are defined as plant exudate produced by plants either naturally or when the plant is exposed to physiological damage as a result of a pathogenic condition or mechanical damage as a result of the influence of environmental factors or pest infestation. Resins can be made synthetically by freezing formaldehyde or freezing the resin after mixing it with glycerin such as Colophony resin. In general, the most common plant families that produce resins are Pinaceae, Cupressaceae, Araucariaceae, and Podocarpaceae [3, 4, 9]. The benefits of resins are their use in therapeutic recipes in eastern civilizations, especially in treating burns and superficial and deep wounds, such as Balsam resin, as well as its use in religious rituals, weddings, and astrology. Resins are also used in the manufacture of incense, such as Amber resin, soaps, and cosmetics, such as Myrrh resin. Scientific research has proven that resins have high anti-microbial, antitumor, anti-inflammatory, and anti-skin perfusion efficacy. Resins are one of the plant's defenses against insects, as some resins are formed when they are absorbed by insects to turn this formed sap into a sticky resin that prevents the insect from moving and then eliminates it, such as Shellac resin. The resins are divided into several groups; the oleo-resin group is composed of the resin and the volatile oil such as Copaiba resin, which includes in its composition diterpenes or sesquiterpenes. The second group is gum-resin which is a mixture of resin and gum-like gamboge. The third group is the oleo-gum resins, which consist of resin, gum, and volatile oil, such as asafoetida resin extracted from the rhizomes of the roots of the plant Ferula asafoetida, which consists of ferulic acid and the compound Umbelliferone and volatile oils such as sesquiterpenes such as foetidine, saradaferin, methoxy courmarin, and polysaccharides. The fourth group is glyco-resin is a mixture of resin and sugar such as jalapin and podophyllin resin. The fifth group is Balsams, which are resinous materials that contain in their composition aromatic acids such as cinnamic acid and benzoic acid or both, or esters of these acids such as Peru balsam, Tolu balsam and Storax balsam, which contain a high percentage of Aromatic balsamic acids [4, 6].

## 3. Plant hormones

Plant hormones are chemical runners that are created in one tissue and regulate cellular actions in another tissue by linking with certain proteins that role as receptors associate to cellular transduction pathways. The plant hormones are synthesized in one tissue and react on specific target sites in another tissue at very low concentrations. Plant hormones that are transferred to sites of activity in tissues far away from their site of biosynthesis are indicated as endocrine hormones. Those that react on cells of tissue close by the source of biosynthesis are indicated as paracrine hormones. Plant growth and development are modulated by six major groups of hormones: auxins, cytokinins, gibberellins, abscisic acid, ethylene, and brassinosteroids [3, 10]. A diversity of other signaling compounds that play roles in impedance to pathogens and protection against herbivores have also been specified in plants, including combined and uncombined forms of jasmonic acid, salicylic acid, and small polypeptides. Another compound, strigolactone, has lately been shown to be an intendable signaling compound that regulates the growth of lateral buds [10, 11]; this compound may also be a valid plant hormone. Other groups of compounds, such as flavonoids, work as both intracellular and extracellular regulators of signal transduction pathways [12]. Indeed, the list of signaling factors and growth regulators continues to expand.

### 3.1 Auxins

The first signaling compound is the hormone auxin. Auxin was the first growth regulator to be calculated in plants, and a lot of the early physiological reports on the mechanism of plant cell extension were executed about auxin action. Auxin signaling has been begun to purpose in nearly every feature of plant growth and development. Moreover, auxin and cytokinin differ from the other plant growth regulators and signaling compounds in one important subject: they are desired for plant embryo viability. Whereas other plant growth regulators seem to work as regulators of separate development processes, auxin and cytokinin seem to be desired at several levels less or more continuously. The various growth and development processes that are controlled by auxin are apical dominance, stem elongation, fruit development, root initiation, oriented or topic growth, and meristem development [3, 10]. The Went's studies with gelatin and agar blocks demonstrated unequivocally that growth-promoting influence diffusing from coleoptile tip was chemical substance. The fact that it was produced at one location and transported in minute amounts to its site of action qualified it as an authentic plant hormone. In the mid-1930s it was determined that the principal natural auxin is indole-3-acetic acid (IAA), (**Figure 4**) [3, 13, 14].

### 3.2 Gibberellins

The second group of plant growth regulators to be recognized is the gibberellins (GAs). At least 136 natural types of GAs was produced in plants have been

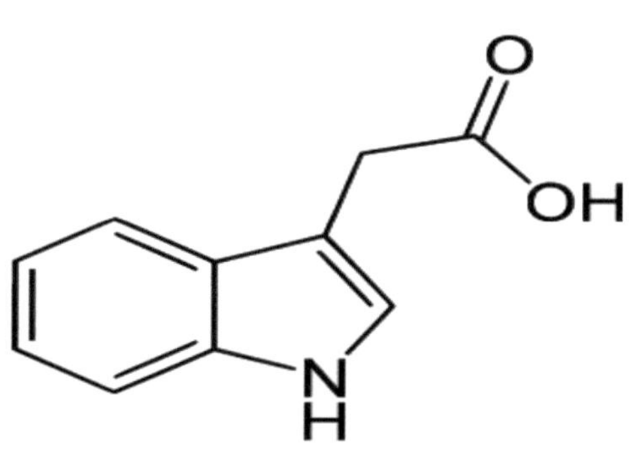


**Figure 4.**
*Indole-3-acetic acid (IAA) structure.*

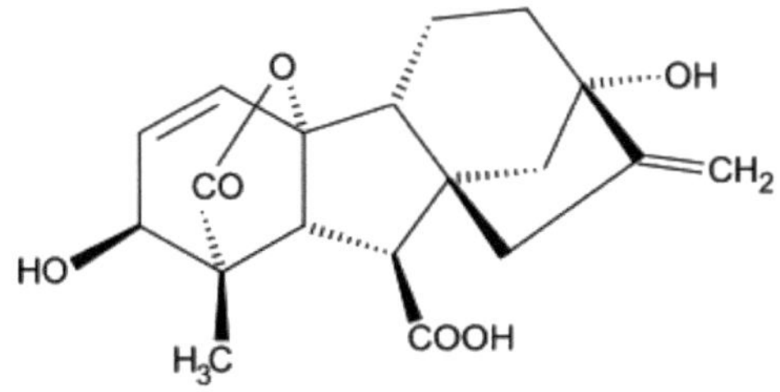


**Figure 5.**
*Gibberellin (GA) structure.*

identified [15]. Opposite of the auxins, which are identified by their biological characters, the gibberellins all share a homogenous chemical structure but relatively few of them have essential biological activity (**Figure 5**). Many of the gibberellins that do not have base biological activity are either precursor compounds of the bioactive gibberellins or their destruction products. Gibberellins also play main roles in a variety of other physiological processes, such as the transition to flowering, seed germination, and pollen growth and development. The biosynthesis of gibberellins is under rigid genetic, environmental, and developmental control [13, 14]. Gibberellins are best known for their enhancement of stem elongation, and gibberellin-deficient mutants that have dwarf phenotypes have been separated. Gibberellins first came to the observation of Western scientists in the 1950s; they had been discovered much earlier in Japan. Rice farmers had long known the fungal disease termed 'foolish seedling' that caused rice plants to grow too tall and discarded seed production. The pathologists of plants found that these symptoms of infection in rice were caused by *Gibberella fujikuroi*, which had infected the plants. The cultures of this fungus and chemical analysis in the laboratory enabled Japanese scientists in the 1930s to obtain impure substances with plant growth-enhancing activity. They named this combination of compounds gibberellin A. Gibberellins to represent a large family of tetracyclic diterpene acids biosynthesized via a terpenoid pathway. Knowledge of gibberellin synthesis and deactivation is important to understanding gibberellin homeostasis. Homeostasis depends upon the regulation of gibberellin biosynthesis, transport, and deactivation [3, 14].

### 3.3 Cytokinins

The cytokinins are reverses of auxins, being biosynthesized in roots but with the most spectacular effects on shoot formation. However, shoot tissues can also synthesize cytokinins, as can germinate seeds. A traditional example of cytokinins is coconut milk, the profuse liquid endosperm of the coconut fruit, which is still a common cytokinin origin in the plant cell, tissue, and organ culture media. Cytokinins were at first named for their ability to stimulate cell division, but they also purpose in the induction of shoots, retardation of senescence, and dormancy release [3, 16]. Cytokinins are imitative of adenine, one of the purine bases create in all RNA and DNA. The four main groups of natural cytokinins each have a different five-carbon side-chain linked to the N6 position. The major free cytokinin groups, dihydro-zeatin and trans-zeatin are more biologically active than the two groups found in tRNA (isopentenyl adenine and cis-zeatin) [13, 16]. The side chains of naturally revolving cytokinins are chemically related to carotenoid pigments, rubber, the plant hormones abscisic acid and gibberellin, and the plant defense substances known as phytoalexins. All of these compounds are created from isoprene units. Isoprene is alike in structure to the side chains of iP and zeatin. These cytokinin side chains are biosynthesized from isoprene imitative. Large molecules

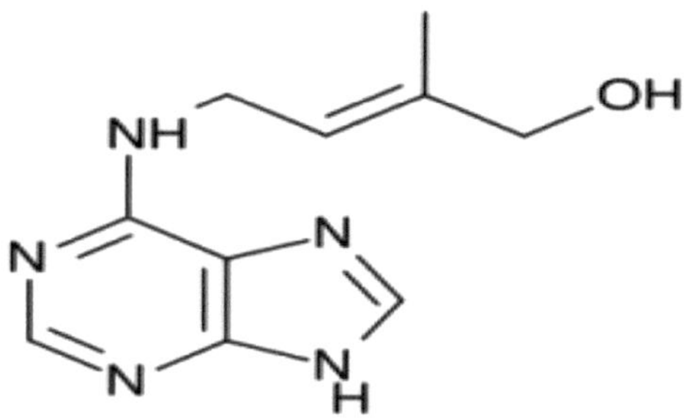


**Figure 6.**
*Cytokininin (Zeatin) structure.*

of the carotenoids and rubber are initiated by the polymerization of many isoprene units; cytokinins consist of fair one of these units. The precursor for the initiation of these isoprene units in cytokinins is dimethylallyl diphosphate (DMAPP), which is derived from either the methylerythritol phosphate (MEP) pathway (primary for DHZ, trans-zeatin, and iP) or the mevalonate pathway (primary for ciz-zeatin) (**Figure 6**) [3].

### 3.4 Abscisic acid

Abscisic acid (ABA) is a growth retardant name because this hormone is related to abscission layer formation (**Figure 7**). ABA does promote fruit drop, growth retardant, and closing stomata in plant leaves [17, 18]. ABA is a 15-carbon molecule and its biosynthesis occurs from the malfunction of carotenoid pigments, especially violaxanthin, a 40-carbon molecule. Formerly, mevalonic acid was believed to be the major precursor, with soon steps in similar with gibberellin biosynthesis. This other pathway may utilize in tissues such as in tomato seedlings and avocado mesocarp [19, 20]. ABA is synthesized in large quantities in water-stressed plant tissues, especially leaves and roots, but also has a role in seed ripening, senescence, and dormancy. ABA concentrations are decreased by oxidative suppression to phaseic acid or by the synthesis of glucosides [3, 13].

### 3.5 Ethylene

Ethylene ($C2H4$) is a unique gaseous hormone that diffuses rapidly out of plant tissues. Its direct precursor is 1-aminocyclopropane-1-carboxylate (ACC) which in turn produced from S-adenosyl methionine, an imitative of another common amino acid (methionine). Ethylene is synthesized in response to cell injury and other stresses such as deficient oxygen (**Figure 8**). It cumulates rapidly during fruit ripening and senescence stages, but all living cells synthesize ethylene. Oxidation and conjugation can happen, but dispersion into the atmosphere is probably the main elimination pathway [3, 13, 21].

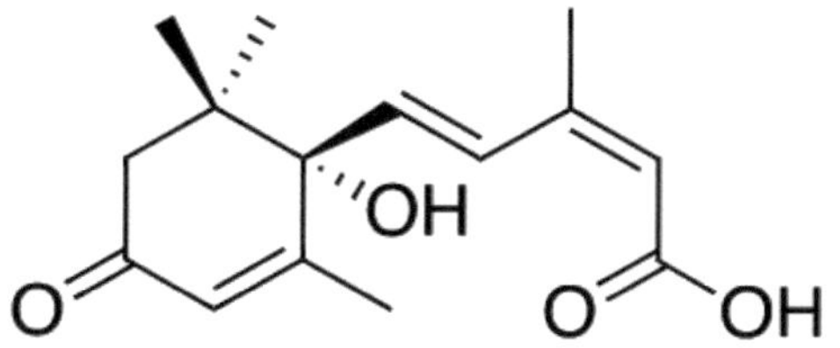


**Figure 7.**
*Abscisic acid (ABA) structure.*

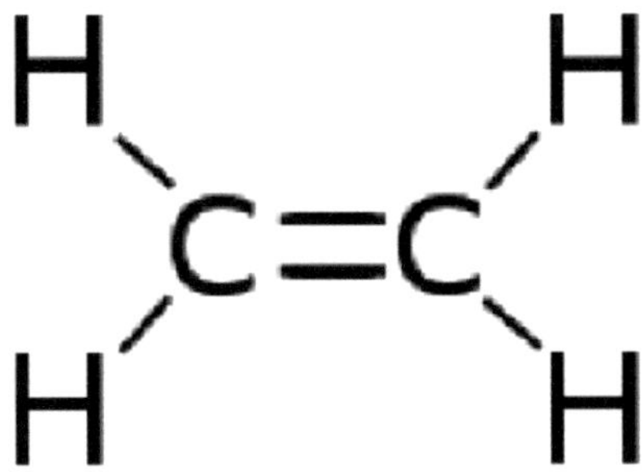


**Figure 8.**
*Ethylene ($C_2H_4$) structure.*

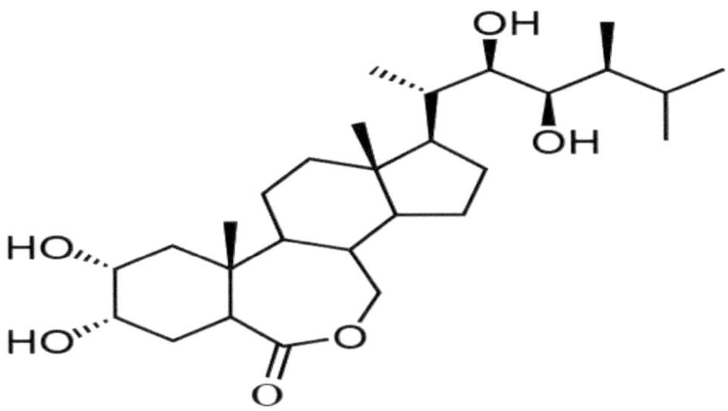


**Figure 9.**
*Brassinosteroid (BR) structure.*

### 3.6 Brassinosteroid

Steroid hormones have extended been recognized in animals, but they have only lately been revealed in plants. Animal steroid hormones involve the sex hormones (androgens, estrogens, and progestins) and the adrenal cortex hormones (mineralocorticoids and glucocorticoids). The brassinosteroids (BRs) are a class of steroid hormones that play more important roles in a wide domain of developmental processes in plants, including cell division and elongation in roots and stems, reproductive development, photomorphogenesis, stress responses, and leaf senescence (**Figure 9**) [22]. Studies by Mitchell et al. [23] showed that the utmost growth-promoting activity was found in the organic extract of pollen from the rape plant (*Brassica napus* L.). Such as abscisic acid and gibberellin, brassinosteroids are biosynthesized as a section of two farnesyl diphosphates to produce the C30 triterpene squalene. Squalene then succumbs to a series of ring closures to produce the pentacyclic triterpenoid (sterol) precursor cycloartenol. All steroids in plants are obtained from cycloartenol by other modifications and a series of oxidation reactions [3, 24].

## 4. Production of secondary metabolites in plant tissue cultures

Tissue cultures of plants are used to produce large quantities of secondary metabolic products, although cultures of callus and cell suspensions often do not produce higher levels of the whole plant. Therefore, some technologies were used to increase the production of secondary metabolites by plant tissue culture techniques through the selection of high-production cells. This is done after separating the high-production cells from their low-production counterparts, and the latter are usually excluded by visual methods [2]. The separation process of produced cells from others is carried out using cell cloning technology, which is an easy and simple method in which single cells are taken from mostly cell suspensions that are cultured on a suitable medium. After the formation of cell masses from single cells, each cell mass is sieved separately and the types and quantities of secondary metabolites it contains are determined. The process of selecting high-producing plant cells for secondary metabolites begins with the selection of a plant with a high production for the desired secondary compound or compounds by selecting the suitable explant, it's surface sterilization, and in vitro culture on a medium prepared for the initiation of callus cells. Then the formed callus masses are culture in cell suspension cultures, from which the inoculums are transferred and spread on a solid medium [2, 25, 26].

## 5. The effect of the components of the medium on the production of secondary metabolites

The growth of plant cells in tissue cultures occurs when the requirements for division and growth are available for them from nutrients, growth regulators, and any other additives that all affect the metabolic activities within the cells. To achieve optimal productivity of secondary metabolites, it is preferable to produce cells in a medium that is optimal for increasing biomass. Then the cells are transferred to the production medium that achieves the highest yield of the desired compound. Note that it is not necessary for the callus medium or the perpetuation medium to be ideal for the production of secondary products. Therefore, many growth regulators and other additives are being tested to obtain an optimal medium for production. The components of the nutrient medium in general, such as carbon source, nitrogen, phosphate, growth regulators, precursors, stimulants, vitamins, additives, and others, affect the fluctuation of the production of secondary metabolites [2, 26, 27].

### 5.1 Effect of carbon source

The carbon source generally affects the production of secondary metabolic compounds. For example, an increase in sucrose in the production medium from 4 to 10% led to an increase in the production of alkaloids in tissue cultures of *Catharanthus roseus*. It was also found that the addition of sucrose as a carbon source was better than fructose and galactose when producing diosgenin from tissue cultures of *Dioscorea deltoidea* and *Dalanites aegyptica*. An increase in Ubiquinone-10 was also recorded in tobacco tissue cultures when low levels of sucrose were added to the production medium. In another study, it was found that adding 40 g $L^{-1}$ of sucrose to the medium in tissue cultures in the dark led to an increase in the accumulation of the proanthocyanin compound in the plant *Hypericum perforatum*. While the accumulation of kaempferol compound when adding 50 g $L^{-1}$ sucrose to the medium in tissue cultures of the same plant exposed to light [2].

### 5.2 The effect of a nitrogen source

Adding high concentrations of nitrogen sources to the media in tissue cultures stimulates cells to synthesize amino acids and proteins, including enzymes and nucleic acids. The primary products of metabolism contain nitrogen, which directly affects the formation of secondary metabolic products. In general, high concentrations of nitrogen added to the medium lead to inhibition of the synthesis of secondary metabolites. The addition of potassium nitrate and ammonium nitrate in high concentrations to the medium prepared for tissue cultures leads to inhibition of the production of anthocyanins by 90% and alkaloids by 80% [2].

### 5.3 Effect of phosphate

Many secondary metabolites are produced from phosphorylated intermediates, which in turn release phosphate. Inorganic phosphates are essential in photosynthesis and respiration. Generally, high levels of phosphate stimulate cells to divide, grow, and synthesize primary metabolites. When the concentration of phosphate in the tissue cultures increases, it leads to an increase in the production of alkaloids in the plant *C. roseus*, the anthraquinone compound in the *Morinda citrifolia* plant, and the diosgenin compound in the *D. deltoidea* plant. Other studies found that decreasing the concentration of phosphate in tissue cultures led to an increase in anthocyanins and phenols in *C. roseus* and an increase in alkaloids and solasodine in *Solanum laciniatum*. While increasing or decreasing the phosphate concentration did not affect the production of the protoberberine alkaloid in the tissue cultures of *Berberis* sp. [2].

### 5.4 Effect of precursors

Precursors are called substrate molecules that can be incorporated into secondary metabolites and added to the medium prepared to produce the desired secondary compounds. In general, the addition of precursors stimulates the production of secondary metabolites, although it inhibits the growth of tissue cultures in several cases. For example, the addition of precursors to the medium prepared for tissue cultures of the *Datura* spp. plant led to a noticeable increase in the production of alkaloids, but this was occurred opposite by inhibition in the growth of cultures after the addition of ornithine, phenylalanine, tyrosine, or sodium phenylpyruvate. It was proven that there was a significant increase in the accumulation of ajmalicine in the cultures of the callus of *Coleus blumei* plant when the medium was enriched with the precursor tryptamine and rosmarininc acid accumulation when the medium was enriched with tryptamine compound and 50 g $L^{-1}$ sucrose. Also, rosmarinic acid accumulated in high concentrations in the stem segment explants of the same plant when the liquid medium was included with a concentration of 10 or 20 mg $L^{-1}$ of proline acid [2, 28].

### 5.5 Effect of plant growth regulators

Plant growth regulators such as auxins and cytokinins affect cell division, various metabolic processes, and plant growth in tissue cultures. Several scientific articles indicated that the type of growth regulator and its concentration affected the productivity of tissue cultures from secondary metabolites. It was found that the addition of auxin indole acetic acid, indole pyruvic acid, or naphthalene acetic acid to the medium prepared for tissue cultures of *Balanites aegyptica* increased the production of diosgenin. The addition of auxin in some cases inhibited the production of some secondary metabolites, such as inhibiting anthocyanin synthesis in carrot plant tissue cultures after enhancing the medium with naphthalene acetic acid and indole acetic acid [2, 29]. The addition of auxin 2,4-dichlorophenoxy acetic acid (2,4-D) to the tissue cultures of tobacco led to the inhibition of the production of alkaloids as well as shikonin in the tissue cultures of *Lithospermum erythrorshizon*. Another study also found that the addition of cytokinins to the cultures of

*C. roseus* stimulated the production of ajmalicine, and the tissue cultures of tobacco led to the production of scopolamine and scopoletin compounds, and the tissue cultures of *Ricinus* sp. led to the production of carotenoids. Cytokinins inhibited the production of secondary metabolites in some tissue cultures, such as anthropoquinines in *M. citrifolia* plant cultures, nicotine in tobacco plant cultures, and Chicoine in *Lithospermum erythrorshizon* cultures. It is noted from previous studies that the addition of auxins and cytokinins to the medium separately did not give positive results in stimulating the production of secondary metabolic compounds in most cases. In general, many studies showed that the addition of combinations of auxins and cytokinins stimulated the increase in the production of secondary metabolites [2, 26].

## 6. Effect of some growth regulators on the production of secondary metabolites in some medicinal plants

### 6.1 *Stevia rebaudiana*

The leaves of shoots that cultured on Woody Plant Medium (WPM) supplemented with 2.27 mM thidiazuron (TDZ), 4.54 mM TDZ, 2.22 mM benzyl adenine (BA) + 2.69 mM naphthalene acetic acid (NAA), 2.22 mM BA +5.37 mM NAA, 2.32 mM kinetin (Kn) +5.71 mM indole acetic acid (IAA), or 2.32 mM Kn + 2.69 mM NAA led to stimulate steviolbioside, rubusoside, and dulcoside compounds by in vitro culture technique [30].

The 4.6 pH of the medium was the main factor for increasing concentrations of secondary metabolite compounds in stevia leaves by in vitro culture technique. The phenols and flavonoids were increased when cultured on a medium supplied with the combination of BA and GA3 or IAA compared to separately applied growth regulators appearing synergistic effects of plant growth regulators (especially of auxins and cytokinins). A positive correlation was found between the flavonoids, phenols, and the antioxidant activity in the *S. rebaudiana* extracts [31].

The highest callus-induction frequency and callus-mass increase were obtained from MS medium supplemented with 2.0 μM NAA. The leaf explants that cultured on MS medium supplemented with 2.0 μM NAA led to the highest concentration of steviol glycosides, flavonoids, and phenols, and higher antioxidant activity was determined in the secondary metabolite compounds of callus from leaf segments. Proline acid reduced the concentration of flavonoids and steviol glycosides. The callus from leaf explants that cultured on MS medium supplemented with 2.0 μM NAA and 2.0 μM proline acid recorded the highest concentration of total phenolic compounds [32].

### 6.2 Pimpinella alpeno

The results of one study showed that adding 200 mg $L^{-1}$ IAA and 25 mg $L^{-1}$ gibberellic acid ($GA_3$) to the medium prepared for tissue cultures of *Pimpinella alpino* leaves increased the production of saponin. While when adding 100 mg $L^{-1}$ IAA and 25 mg $L^{-1}$ $GA_3$ to the medium, it led to a decrease in the production of saponin compound in the leaves and it reached the lowest value [33]. $GA_3$ affects metabolism and nucleic acid which plays an important role in protein biosynthesis and enhanced the activity of enzymes for plant growth and development. Increased protein biosynthesis as crude material essential enzymes in plant metabolism and increase the production of the secondary metabolite compounds, including saponins at the final stages [34].

### 6.3 *Crysanthemum cinerariefolium*

The leaf segments of *Crysanthemum cinerariefolium* plant that cultured on MS medium supplemented with 4 mg $L^{-1}$ 2, 4-D and 0 mg $L^{-1}$ kinetin recorded the best-produced callus by *in vitro* culture technique. The callus contains the secondary metabolite compounds such as some of the flavonoid quercetin precursors such as tetrahydroxy chalcone and acetic acid and some other secondary products [35].

## 7. Conclusions

* There are factors that affect the increase in the induction and production of secondary metabolites from plants that can be applied and utilized in extracting effective compounds from medicinal plants that are used in the industry of medicines and pharmaceuticals.

* The levels of bioactive compounds in medicinal plants vary depending on the type of plant tissue.

* The possibility of using the plant tissue culture technique in the production of secondary metabolites from the explants of medicinal plants.

* Increasing the concentrations of plant growth regulators such as auxins or cytokinins or adding them in ideal combinations leads to an increase in the induction of secondary metabolites in tissue cultures of medicinal plants.


## Author details

Majid Ibrahim
College of Agriculture, University of Basrah, Basrah, Iraq

*Address all correspondence to: majid.abdulhameedl@uobasrah.edu.iq

## References

[1] Dewick PM. Medicinal Natural Products: A Biosynthetic Approach. 2nd Edition. New York: John Wiley & Sons; 2002

[2] Ibrahim KM. Applications of Plant Biotechnology. Iraq: Universal House to Printing, University of Al-Nahrain, Ministry of Higher Education and Scientific Research; 2017

[3] Taiz L, Zeiger E. Plant Physiology. 5th ed. Sunderland. MA: Sinauer Associates; 2010

[4] Alasady MHS. Basics of Medicinal Plants and their Active Constituents. Baghdad, Iraq: House of Books and Documents Printing; 2018

[5] Van Wyk BE, Wink M. Medicinal Plants of the World. Germany: CABI; 2018

[6] Hussein RA, El-Anssary AA. Plants secondary metabolites: The key drivers of the pharmacological actions of medicinal plants. Herbal Medicine. 2019;**1**:13

[7] Li Y, Kong D, Fu Y, Sussman MR, Wu H. The effect of developmental and environmental factors on secondary metabolites in medicinal plants. Plant Physiology and Biochemistry. 2020;**148**:80-89

[8] Namdeo AG. Plant cell elicitation for production of secondary metabolites: A review. Pharmacognosy Reviews. 2007;**1**(1):69-79

[9] Bhumi G, Savithramma N. Screening of pivotal medicinal plants for qualitative and quantitative phytochemical constituents. International Journal of Pharmacy and Pharmaceutical Sciences. 2014;**6**(3):63-65

[10] Davies PJ, editor. Plant Hormones and their Role in Plant Growth and Development. Dordrecht, Netherlands: Springer Science & Business Media; 2012

[11] Brewer PB, Dun EA, Ferguson BJ, Rameau C, Beveridge CA. Strigolactone acts downstream of auxin to regulate bud outgrowth in pea and Arabidopsis. Plant Physiology. 2009;**150**(1):482-493

[12] Peer WA, Murphy AS. Flavonoids and auxin transport: Modulators or regulators? Trends in Plant Science. 2007;**12**(12):556-563

[13] Turnbull CG. Plant hormones: chemical signaling in plant development. In: Atwell BJ, Kriedemann PE, Turnbull CGN, Eamus D, Bieleski RL, Farquhar G, editors. Plants in Action: Adaptation in Nature, Performance in Cultivation. South Yarra, Victoria: MacMillan Education Australia; 1999. pp. 284-305

[14] Davies PJ. The plant hormones: Their nature, occurrence, and functions. In: Plant Hormones. Dordrecht: Springer; 2010. pp. 1-15

[15] MacMillan J. Occurrence of gibberellins in vascular plants, fungi, and bacteria. Journal of Plant Growth Regulation. 2001;**20**(4):387-442

[16] Galuszka P, Spíchal L, Kopečný D, Tarkowski P, Frébortová J, Šebela M, et al. Metabolism of plant hormones Cytokinins and their function in Signaling, cell differentiation and plant development. In: Studies in Natural Products Chemistry. Vol. 34. Amsterdam, Netherlands: Elsevier; 2008. pp. 203-264

[17] Davies PJ, editor. Plant Hormones: Physiology, Biochemistry and Molecular Biology. Ithaca, New York, USA: Springer Science & Business Media; 2013

[18] Nambara E, Marion-Poll A. Abscisic acid biosynthesis and catabolism.

Annual Review of Plant Biology. 2005;**56**:165-185

[19] Milborrow BV. The reduction of (±)-(2– 14C) Abscisic acid to the 1′, 4′-trans-diol by pea seedlings and the formation of 4'-Desoxy ABA as an artefact. Journal of Experimental Botany. 1983;**34**(3):303-308

[20] Willows RD, Netting AG, Milborrow BV. Endogenous biosynthetic precursors of (+)-abscisic acid. I. Incorporation of isotopes from 2H2O, 18O2 and [5-18O] mevalonic acid. Functional Plant Biology. 1994;**21**(3): 327-343

[21] Robles L, Stepanova A, Alonso J. Molecular mechanisms of ethylene–auxin interaction. Molecular Plant. 2013;**6**(6):1734-1737

[22] Clouse SD, Sasse JM. Brassinosteroids: Essential regulators of plant growth and development. Annual Review of Plant Biology. 1998;**49**(1):427-451

[23] Mitchell JW, Mandava N, Worley JF, Plimmer JR, Smith MV. Brassins—A new family of plant hormones from rape pollen. Nature. 1970;**225**(5237): 1065-1066

[24] Muthulakshmi S, Pandiyarajan V. Influence of Brassinosteroids (BRs) on the vincristine content of *Catharanthus roseus* (L.) G. Don. European Journal of Experimental Biology. 2015;**5**(10):54-56

[25] Yamamoto H, YATo A, Yazaki K, Hayashi H, Taguchi G, Inoue K. Increases of secondary metabolite production in various plant cell cultures by co-cultivation with cork. Bioscience, Biotechnology, and Biochemistry. 2001;**65**(4):853-860

[26] Smetanska I. Production of secondary metabolites using plant cell cultures. Food Biotechnology. 2008;**111**:187-228

[27] Ahsan T, Chen J, Wu Y, Irfan M. Application of response surface methodology for optimization of medium components for the production of secondary metabolites by *Streptomyces diastatochromogenes* KX852460. AMB Express. 2017;**7**(1):1-10

[28] Namdeo AG. Plant cell elicitation for production of secondary metabolites: A review. Pharmacognosy Reviews. 2007;**1**(1):69-79

[29] Khan N, Bano A, Zandi P. Effects of exogenously applied plant growth regulators in combination with PGPR on the physiology and root growth of chickpea (*Cicer arietinum*) and their role in drought tolerance. Journal of Plant Interactions. 2018;**13**(1):239-247

[30] Röck-Okuyucu B, Bayraktar M, Akgun IH, Gurel A. Plant growth regulator effects on in vitro propagation and stevioside production in *Stevia rebaudiana* Bertoni. HortScience. 2016;**51**(12):1573-1580

[31] Radić S, Vujčić V, Glogoški M, Radić-Stojković M. Influence of pH and plant growth regulators on secondary metabolite production and antioxidant activity of *Stevia rebaudiana* (Bert). Periodicum Biologorum. 2016;**118**(1):9-19

[32] Blinstrubienė A, Burbulis N, Juškevičiūtė N, Vaitkevičienė N, Žūkienė R. Effect of growth regulators on *Stevia rebaudiana* Bertoni callus genesis and influence of auxin and proline to steviol glycosides, phenols, flavonoids accumulation, and antioxidant activity *In vitro*. Molecules. 2020;**25**(12):2759

[33] Fathonah D, SUGIYARTO S. Effect of IAA and GA3 toward the growing and saponin content of purwaceng (*Pimpinella alpina*). Nusantara Bioscience. 2009;**1**(1):17-22

[34] Martin R, editor. Protein Synthesis: Methods and Protocols. Totowa, NJ: Humana Press; 1998

[35] Purwianingsih W, Febri S, Kusdianti. Formation flavonoid secondary metabolites in callus culture of *Chrysanthemum cinerariefolium* as alternative provision medicine. In: AIP Conference Proceedings (Vol. 1708, No. 1). Bandung, Indonesia: AIP Publishing LLC.; 2016. p. 030005. DOI: 10.1063/1.4941150

Chapter 9

# Responses of Agronomically Important Tropical Crops to the Application of Brassinosteroid

*Juan F. Aguirre-Medina, Jorge Cadena-Iñiguez and Juan F. Aguirre-Cadena*

## Abstract

Brassinosteroids (Br) have been shown to favor the growth and reproduction of crops under adverse environmental conditions, which negatively affect their growth and production. In order to solve some of the problems in the field with various perennial crops, the application of a homobrassinolide (HBr) (CIDEF-4) has been investigated under *in vitro* and *ex vitro* conditions to evaluate growth at different concentrations in *Musa* spp. L. and *Saccharum officinarum* L and in the field with foliar applications in *Theobroma cacao* L, *Mangifera indica* L and *Coffea arabica* L. to evaluate yield and quality of fruits. Morphological and physiological yield components were recorded. The results indicate in the *in vitro* evaluations, increased regrowth height and *ex vitro* differences in growth are improved by increasing the number of applications. In cocoa and coffee plants, flowering and yield are influenced. The high concentrations applied did not necessarily increase the crop yield or the combination with potassium nitrate. In mango, the quality of the fruits was better when applying the HBr alone or in combination with nitrate in fruit firmness, and total soluble solids improved.

**Keywords:** brassinosteroids, growth, quality fruits, yield, biomass

## 1. Introduction

The brassinoesteroides (Br) are the most active growth-promoting extracts isolated from *Brassica napus* pollen [1]. These are steroidal phytohormones essential for the growth, development and adaptation of plants to biotic and abiotic stress [2–5]. Regulatory activity seems to be due to the influence of these hormones on metabolic processes related to photosynthesis, nucleic acid and protein biosynthesis [6, 7]. It is transported from sites of synthesis to sites of distant action and operates at very low concentrations [8–10]. These compounds have a wide range of activity in protecting plant metabolism when these are under stress conditions and consequently increase their yield [11], through de increase la activity within the antioxidant route [12, 13], promote membrane polarization and influence tolerance to different stressful environmental conditions [14–17]. These are low and high temperatures [18, 19] and tolerance to water stress [20] among others. At the cellular level, they are expressed in elongation, cell differentiation and genetic modulation [7, 21].



IntechOpen

In agricultural activities, it has been shown that various abiotic stressors for crops can be mitigated through the application of hormones, such as brassinosteroids [6]. This technology is relevant when considering that agricultural activities each year face environmental factors, such as drought and erratic rainfall [22–24], which has been the main cause of food insecurity that induces decreased yields [25]. Nowadays some Brassinosteroids homologos (HBr) which are similar to natural brassinosteroids, have been shown to be useful in agriculture [26]. They have been used as a strategy to increase growth and yield in various crops in the field and greenhouse through their exogenous application, and these have shown enhancement of growth and reproduction under adverse conditions [11, 16, 27–29]. In addition, HBRs play a fundamental role in flower and fruit development, leaf senescence and abscission [30, 31]. They are also proposed in the reproduction of plants *in vitro* to be added to the culture medium in different explants and organs [32] as substitutes for auxins and cytokinins [33, 34]. Under these conditions, they can stimulate aerial and radical plant growth [35, 36], as noted in rooting, number of leaves and plant height in various orchids when added HBr (Cidef-4) to the medium Yasuda [37] and in *Cedrela odorata* L. the number of shoots and leaves is increased by adding the same (Hbr) to the Murashige and Skoog medium (MS) [38, 39].

In the greenhouse, biomass, stevioside, rebaudioside and steviol content are increased in the *Stevia* crop compared to the control [40]. In tomato, it is associated with the content of lycopene and carbohydrates [41] and in transgenic tomatoes, early flowering and a slight increase in fruit yield and number of fruits per plant were obtained [20, 42], fruits and oilseeds [43–45] in addition to enhancing crop quality [46].

## 2. Response of agricultural crops to the application of brassinosteroids through the increased lab activity

### 2.1 *In vitro* reproduction of *Musa* spp. "big dwarf banana" with brassinosteroid and other growth regulators

Banana (*Musa* spp.) is widely consumed by the world population and in tropical regions. It is a daily food in developing countries and its *in vitro* reproduction often has a low multiplication coefficient, high percentage of phenolization in explants and low survival [47, 48]. In other cases, abnormal shoots are reported in plants with some growth regulators, which are traditionally incorporated into the growing medium [33].

In this regard, non-traditional growth regulators, such as HBr, can favor multiplication processes [18, 49] and improve normal plant growth and development. However, by subjecting the explant to an artificial environment for its reproduction. It is exposed to stressful conditions, which can be attenuated by HBr and improve its reproduction, as happened with the FHIA-21 banana, which favored the formation of roots *in vitro* [50].

The *in vitro* response of *Musa* spp. cv great dwarf shows morphological changes when establishing the apical meristem in MS medium [38] and three growth regulators, 6-bencilaminopurine (BAP), indoleacetic acid (IAA) and homobrasinolide (HBr-Cidef-4; Natura del desierto SA de CV, México) in two concentrations 2 and 4 $mg^{L-1}$, during 80 days of evaluation with changes of medium every 20 days.

The HBr, in its two concentrations, presents a response similar in the number of leaves and roots to those induced with BAP and IAA and statistically different from the control ($P \leq 0.05$) (**Table 1**).

The lowest average shoot height was with BAP at the lowest concentration. On the other hand, by increasing the concentration, the height was almost doubled.

| Treatment ($mgL^{-1}$) | Height ($cm.regrowth^{-1}$) | Number of $leaves.regrowth^{-1}$ | Number of roots |
|---|---|---|---|
| BAP 2.0 | 3.59 e | 11.25 a | 5.12 b |
| BAP 4.0 | 6.07 bc | 11.12 a | 7.37 a |
| AIA 2.0 | 4.24 de | 10.75 a | 8.12 a |
| AIA 4.0 | 6.07 bc | 9.87 ab | 6.50 ab |
| HBr 2.0 | 7.48 a | 10.75 a | 7.75 a |
| HBr 4.0 | 7.00 ab | 10.87 a | 7.50 a |
| Control | 5.35 bc | 7.62 b | 6.50 ab |
| CV (%) | 13.4 | 16.4 | 18.4 |

**Values with different letters within each factor and column are statistically different ($p \leq 0.05$). BAP(6-benzylaminopurine), HBr (Brassinosteroid Cidef-4), IAA (Indolacetic Acid).*

**Table 1.**
*Morphological components in apical meristem explant of* Musa *spp. cv gran enano in interaction with different concentrations of growth regulators during the* in vitro *multiplication stage.*

An increase in regrowth height from 0.5 mgL of BAP has been reported [51]. This cytokinin has been widely used in the *in vitro* regeneration of various plants [52] and induces the formation of axillary shoots and adventitious roots of the meristematic explant in banana [53] and their efficacy in different banana cultivars has been demonstrated in foliar growth [54]. Auxins in *in vitro* culture induce root formation and shoot growth [54, 55]. In the case of HBr, the response was different, that is, the inclusion of the two concentrations in the medium induced a similar increase in the height of the explant and was superior to the other treatments ($P \leq 0.05$) [56]. Stems were thicker with HBr and thin with narrow leaves in the control. In *Solanum tuberosum* L. the height of the explants was also increased after 30 days of age, by adding 1 mgL-1 of HBr Biobrás-6 to the medium.

Likewise, pronounced effect on the stimulation of proliferation and shoot length when HBr (0.2 μM) was added to the MS medium compared to the banana control [57]. Robres-Torres [58] add that the application of brassinolide plus benzyladenine (BA) in *Rubus ideaeus* L. did not express dependence in the combination of both regulators in the expression of regrowth, but when applying HBr alone, regrowth elongation was increased.

The number of leaves is higher when including growth regulators compared to the control and they were statistically different ($p \leq 0.5$). The inclusion of HBr increased the leaf area of *Lactuca sativa* L. lettuce with the foliar spraying of HBr Biobras-16 [59] and with the regeneration of adventitious shoots in cauliflower hypocotyls and *Spartina* [60].

When comparing the induction of the number of roots between the three growth regulators, a differential response was found according to the concentration used and without difference with the application of the concentrations in HBr (Cidef-4).

Applying HBr to plants increases the number of roots. Similar result cites [61] when doing repeated foliar sprays with BB-16 (0.05 mgL-1) in *Cattleya leuddemanniana* and *Guarianthe skinneri* (Bateman), Dressler and W. E. Higgins. This same result was obtained with con different orchids, like *Oncidium sphacelatum* Lindl., *Trichocentrum andreanum* (Cogn.) R. Jiménez and Carnevali, *Epidendrum stamfordianum* Bateman, *Guarianthe skinneri* (Bateman), Dressler and W. E. Higgins, *Guarianthe aurantiaca* (Bateman ex Lindl.) Dressler and W. E. Higgins and *Brassavola nodosa* (L.) Lindley, in Chiapas, México [37].

## 2.2 Influence of brassinoesteroid in the acclimation of banana clone *Musa* spp. and *Saccharum officinarum*

The production of micropropagated plantlets is used nowadays to increase new cultivars [62], but there are certain drawbacks in its *ex vitro* phase that limit its widespread use, especially in resisting the stress of transplantation [47, 48]. Exogenous application of HBr has been shown in some plants to improve *ex vitro* survival [18].

The application of HBr in its different frequencies of HBr (Cidef-4) with 4 $mgL^{-1}$, (6 every 14 days, 3 every 28 days and 2 every 42 days), were efficient in inducing the growth of *Musa* spp. However, the more frequent applications induced greater height.

In the case of plant height with the application of HBr every 14 days, it presents a statistical difference ($p \leq 0.05$) with the rest of the treatments during the four samplings and in all the application frequencies it was superior to the control (**Table 2**). The results suggest an interaction between the concentration and its frequency of application, and Mandava [63] adds that the brassinosteroid is more sensitive when the young

| Time (days) | Treatments | Height (cm) | Stem thickness (cm) | Number of leaves |
|---|---|---|---|---|
| 14 | 1. Control | 5.53 b* | 0.53 a | 5.5 ab |
| | 2. HBr every 14 days | 6.41 a | 0.50 a | 5.5 b |
| | 3. HBr every 28 days | 5.95 ab | 0.45 a | 4.8 b |
| | 4. HBr every 42 days | 5.60 b | 0.51 a | 6.0 a |
| | **CV (%) | 8.59 | 10.64 | 11.46 |
| 42 | 1. Control | 10.33 b | 0.63 b | 6.5 a |
| | 2. HBr every 14 days | 12.63 a | 0.65 b | 6.5 a |
| | 3. HBr every 28 days | 11.06 ab | 0.61 b | 6.1 a |
| | 4. HBr every 42 days | 11.25 ab | 0.78 a | 7.0 a |
| | CV (%) | 10.62 | 7.07 | 8.25 |
| 70 | 1. Control | 10.78 b | 0.75 b | 7.8 a |
| | 2. HBr every 14 days | 16.33 a | 0.90 a | 8.0 a |
| | 3. HBr every 28 days | 13.71 b | 0.80 ab | 7.6 a |
| | 4. HBr every 42 days | 13.83 b | 0.86 ab | 8.5 a |
| | CV (%) | 7.72 | 8.87 | 6.65 |
| 84 | 1. Control | 15.21 b | 0.81 a | 8.3 a |
| | 2. HBr every 14 days | 19.17 a | 1.01 a | 8.6 a |
| | 3. HBr every 28 days | 16.33 ab | 0.91 a | 8.1 a |
| | 4. HBr every 42 days | 16.08 ab | 1.03 a | 8.8 a |
| | CV (%) | 14.6 | 17.0 | 9.2 |

*[*]Values with different letters within each factor and column are statistically different ($p \leq 0.05$). HBr (Brassinosteroid 4 $mgL^{-1}$ de Cidef-4).*
***CV = Coefficient of variation. The number of applications. The first 14 days only 1 application in treatment 2. Application at 42 days 3 applications in treatment 2, and 1 application treatment 3 and 4. Evaluation at 70 days, 5 applications treatment 2, 3 application treatment 3 and 1 application treatment 4. Evaluation at 84 days, 6 applications treatment 2, 3 application treatment 3 and 2 application treatment 4.*

**Table 2.**
*Morphological components of* Musa *spp. cv "great dwarf" in interaction with different frequencies of HBr application (Cidef-4).*

tissues present higher content of auxins and the expression of the growth of the plants treated with exogenous HBr can be related to the endogenous plant hormones [18].

Similar results, but with a lower proportion of plant height induction (2%) have been reported by Jeyakumar et al. [64] in "robusta" banana when applying a foliar HBr at a concentration of 0.2 mg $kg^{-1}$ at four and 6 months after having been planted. On the other hand, Izquierdo et al. [33] mention an increase in plant height in banana clone FHI-18 applying by immersion and foliarly before transplantation the HBr BB-6. However, Herrera et al. [65] indicate an increase in the height of plants in banana cultivation with the application of Br applied every 28 days, compared to the application of every 14 days, which, in our case, was the most frequent application and was expressed in the greater height of the regrowth. It is probable that the response is associated with the rapid degradation of Br, as indicated by Janeczko et al. [44], and the more frequent application, may favor its action for a longer time. Application of BR by spraying at specific developmental stages can enhance crop yield [66].

In our case, the thickness of the stem and the number of leaves are the variables with the lowest morphological expression in the different treatments evaluated and they only present statistical differences in two or three of the samplings carried out. On the other hand, Izquierdo et al. [33] reported an increase in the number of leaves and diameter of the pseudostem in banana with the application of HBr, as in the clone FHIA-18, when applying the HBr BB-6.

In potato plants (*S. tuberosum* L.) they found an increase in stem length, fresh biomass and a greater number of minitubers per plant [50].

In the case of the number of leaves, [67] mention a significant increase in the number of leaves of shoots of Vriesea plants (Bromeliaceae) when HBr (MH5) was applied, improving their quality indicators after 49 days of acclimatization. In our case, no differences were expressed between the frequencies of HBr application in this variable.

In the physiological components, the application of HBr alone every 42 days increased the dry biomass of the shoot and was statistically different ($p \leq .05$) to the rest of the treatments, but in the root system the biomass allocation was similar in all the treatments and no statistical difference (**Table 3**).

It has been shown that in the mustard plant *Brassica juncea* (L.) Czern, the yield components were increased with the application of 28-homobrassinolide [43]. Brassinosteroids likely *play a crucial role in modulating plant growth and development, which affect crop architecture and yield* [7].

In *S. officinarum*, plants var. CP 72–2086 were used. The height of the plant with the frequencies and concentrations evaluated of HBr (Cidef-4) was statistically higher than the control ($P \leq 0.05$) (**Table 4**).

| **Treatments** | **(g.plant$^{-1}$)** | |
|---|---|---|
| | **Shoot** | **Root** |
| Control | 2.50 b | 0.62 a |
| HBr every 14 days | 2.65 b | 0.66 a |
| HBr every 28 days | 2.19 b | 0.60 a |
| HBr every 42 days | 3.39 a | 0.66 a |
| **CV (%) | 10.63 | 7.01 |

**Table 3.**
*Comparisons of means of physiological variables of the giant dwarf clone of Musa sp. in the process of acclimatization with different frequencies of application of an HBr at 84 days after sowing.*

| Treatments | Height (cm) | Stem thickness (cm) | Number of leaves |
|---|---|---|---|
| Control | 90.8 b | 6.3 b | 11.5 abc |
| 1% HBr every 14 days (5 applications) | 112.6 a | 7.5 ab | 15.1 a |
| 2% HBr every 14 days | 117.6 a | 7.8 a | 10.0 c |
| 3% HBr every 14 days | 113.6 a | 7.7 a | 12.1 abc |
| 1% HBr every 28 days (2 applications) | 109.8 a | 7.6 a | 11.5 abc |
| 2% HBr every 28 days | 110.8 a | 7.9 a | 10.6 bc |
| 3% HBr every 28 days | 118.6 a | 7.3 ab | 11.5 abc |
| 1% HBr every 42 days (1 application) | 114.0 a | 7.3 ab | 12.3 abc |
| 2% HBr every 42 days | 112.0 a | 7.2 ab | 14.1 ab |
| 3% HBr every 42 days | 117.8 a | 7.3 ab | 11.3 abc |
| **CV (%) | 8.59 | 9.3 | 17.3 |

**Table 4.**
*Morphological components of the yield in* Saccharum officinarum *in interaction with frequencies and concentrations of HBr (Cidef-4).*

In general, the application frequencies in interaction with the concentrations shows a slight increase in plant size as the concentration increased from 1 to 3%. The most frequent applications of 14 and 28 days, also present a certain tendency in the increase of the height of the plant.

HBr (Brassinosteroid Cidef-4), % CV (coefficient of variation). Letters that are not the same indicate statistical difference (Tukey, $P \leq 0.05\%$). Seventy day Data.

In the cane variety (C0. 86,032), brassinolide induced a positive effect on *ex vitro* plants of stem elongation, leaf formation and adventitious shoots during 3 weeks, in addition, survival increased as the concentration of brassinolide increased [68]. The above effect has been demonstrated by inducing elongation and cell division, resulting in growth, thickening and curvature in oat coleoptiles [63].

The applications of exogenous brassinoexteroids induce diverse physiological responses, in addition to cell expansion, vascular differentiation, reproductive development, seed germination, flowering, and fruiting [69, 70].

## 2.3 Flowering and fruiting response of *Theobroma cacao* L

Cocoa was used as food, medicine and tribute in Mesoamerican territory [71, 72]. It is a crop with wide flowering and low fruiting percentage. This is attributed to environmental and hormonal factors.

The application of potassium nitrate alone does not show a positive effect in increasing the flowering and fruiting variables in *T. cacao* L. A similar response is cited by García et al. [73] when applying 2% $KNO_3$ in *Acca sellowiana* [O. Berg] Burret and the lack of response assign it to necrotizing effects on flower buds and fruits. The differences occurred with the application of HBr alone and in combination with different concentrations of potassium nitrate.

The increase in flowers was expressed with the application of HBr alone and was statistically different from the other treatments ($P \leq 0.05$). Combinations with $KNO_3$ present a number of flowers similar to the control (**Table 5**).

*T. cacao* L sowing 3 × 3 with a shade of *Inga micheliana* Harms., *Pouteria sapota* (Jacq.) and *Tabebuia rosea* (Bertol.) A. DC. The values are averages of 5 trees and 4 branches of 0.5 m per tree after the trunk toward each cardinal point.

| Treatments | Number tree$^{-1}$ | | | |
|---|---|---|---|---|
| | Flowers | Opennings Flowers | Cherelles | Fruits |
| 1) Control | 6.5 c** | 17.2 b | 4.32 bc | 1.96 bc |
| 2) 2% de ***$KNO_3$ | 6.3 c | 18.1 b | 3.50 cd | 1.86 bc |
| 3) 1.5% HBr | 9.1 a | 26.5 a | 4.66 b | 1.66 bc |
| 4) 1.5% HBr + 2% $KNO_3$ | 5.7 c | 18.10 b | 3.04 d | 1.32 c |
| 5) 1.5% HBr + 4% $KNO_3$ | 7.8 b | 16.16 b | 4.80 b | 2.42 ab |
| 6) 1.5% HBr + 6% $KNO_3$ | 8.3 ab | 17.46 b | 7.72 a | 2.98 a |
| ****CV % | 7.9 | 7.6 | 10.2 | 22.8 |

*** The letters that are not equal between the columns indicate significant statistical difference ($p \leq 0.05$). ***$KNO_3$ ($13_N$-$2_P$-$44_K$ at the rate of 200 g in water 10 l). HBr was applied only at 1.5% in September and in combination with $KNO_3$ a, 2, 4 and 6% in September, October and November. The values are averages of 15 weeks of evaluation in treatments 1, 2, 3, and 4 that were applied in September. Treatment 5 applied in October with nine samplings and treatment 6 applied in November with 4 samplings. **** Coefficient of variation (%).*

**Table 5.**
*Flowering and fruiting of* Theobroma cacao L. *regional criollo type in response to the applications of potassium nitrate and homobrassinolide.*

The number of flowers increased 40% with the application of HBr and 87% when combined with $KNO_3$ at 2% in relation to the control. It is important to indicate that the application of HBr was in September, the time when the second most important flowering flow begins in *T. cacao* L in this region, and although no detrimental effects were observed in floral structures, the emission was decreased. On the other hand, they increased 27% in relation to the control when combining HBr plus 6% $KNO_3$, which was applied in November. At this date the night temperature decreases and fruiting is favored.

On the other hand, the conversion of flowers to small fruits was higher with the application of $KNO_3$ and HBr in November and this process continued until the formation of fruits, which were statistically different from the rest of the treatments. ($P \leq 0.05$). However, the highest number of fruits found is more related to the date of application of the treatment at the end of the year, that is, October and November. This suggests that the development of the fruit requires the protection of HBr to achieve the conversion of Cherelle to fruit, which seems to be a critical period. In addition to the above, fruiting demands more photosynthates and HBr has been shown to stimulate CO2 assimilation [74, 75]. In other crops, the application of HBr has been effective to promote the increase in yield in annual crops [10, 76] and perennials as in *Passiflora edulis* f. flavicarpa applied after flowering has appeared [77].

## 2.4 Application of a brassinosteroid affects quality in 'ataúlfo' mango

Currently, there is a decrease in mango yield in the south-southeast of Mexico due to various causes attributed to environmental and management factors. Current results in plantations are reflected in high floral abortion rates, premature fruitlet drop and the presence of parthenocarpic fruits or "nubbins. It is considered that HBr can promote fruiting [8, 16, 29] by demonstrating the beneficial effects of HBr in different crops such as vegetables, legumes, cereals, fruits and oilseeds [44, 45, 74], in addition to improving the quality of crops [27, 44, 46, 78] and the yield [46].

Mangoes from all treatments were harvested and the values of each variable were recorded 12 days after harvest (**Table 6**).

| Time (Days) | Treatments | Firmness (N.frut-1) | Total sugars. fruit-1 | Total soluble solids (°Brix.fruit-1) |
|---|---|---|---|---|
| 12 | **$KNO_3$ | 0.61 b* | 0.18 b | 18.05 b |
| | ***HBr(2 g) + $KNO_3$ | 0.64 ab | 0.27 a | 19.80 a |
| | HBr (4 g) + $KNO_3$ | 0.57 b | 0.19 b | 17.92 b |
| | HBr (6 g) + $KNO_3$ | 0.70 a | 0.25 b | 20.07 a |
| | CV % | 5.9 | 7.0 | 3.5 |

**Values with the same letter within each factor and column are equal according to Tukey test P ≤ 0.05. CV = Coefficient of variation, (1) **Control (2% Potassium nitrate KNO3), (2) *** 2 g of HBr Cidef-4/hectare applied on October 16; (3) 4 g of HBr Cidef-4/hectare applied in two equal parts on October 16th and 30th; (4) 6 g of HBr Cidef-4 applied in three equal parts on October 16th and 30th and on November 14th.*

**Table 6.**
*Quality variables during 'Ataúlfo' mango ripening sprayed at preflowering and preharvest with HBr Cidef-4 stored at room temperature (22 ± 2°C).*

The spraying of HBr significantly affected ($P \leq 0.05$) the firmness of the mangoes. The applications of the fruits induced greater firmness in the treatments with 2 g $ha^{-1}$ and 6 g $ha^{-1}$. The greatest difference was registered in the treatment with three applications of HBr. In other crops, such as papayas, they cite an increase in the firmness of the fruits when increasing the application of 24-epibrassinolide (epiBR) to 1 μM [79].

Total sugars increased by combining HBr plus $KNO_3$ (**Table 6**) as the amount of (HBr) increased. In this regard, the highest value was found in the fruits that were sprayed with 2 and 6 g of HBr and it was statistically different from the rest of the treatments ($P \leq 0.05$). In general, the results express a more consistent response in terms of the average sugar content in the treatments applied with potassium nitrate and in the HBr treatments (2 and 6 g $ha^{-1}$).

The amount of TSS found in the treatment with potassium nitrate was 18.0°Brix and in the treatments where HBr was sprayed three times (2 g $ha^{-1}$), the value fluctuated between 19.8° Brix. An increase in the content of total soluble solids without the influence of the concentration of HBr was reported in watermelon [80] but a higher content of total soluble solids is confirmed in *P. edulis* compared to the control with the application of HBr [77]. In papayas, the total soluble solids of the fruits were increased with the application of HBr [79].

In-plant tissues, exogenous applications of HBr induce ethylene production [81, 82] and in some fruits such as mango and strawberries, HBr in small amounts may not be critical for ripening [83, 84], but exogenous applications they are capable of inducing ethylene production [41, 83]. In the Kensington Pride mango variety, ethylene production and peak respiration occurred on the fourth day of ripening, and although the amount of HBr in the fruit was traces, the exogenous application of Epi-BL (45 and 60 ng $g^{-1}$) promoted fruit ripening [83].

## 2.5 Flowering and yield response of *Coffea arabica* cv catimor

*C. arabica* L. plantations currently face contrasting changes in the amount of rainfall and the distribution of rainfall that has decreased their yield. It also interacts with internal, hormonal and nutritional factors [85]. The importance of the interaction of the environment in the flowering of fruit trees has been pointed out [86].

The presence of stressors, biotic or abiotic, that affect crop yield under field conditions, can be attenuated by applying growth regulators, or, when the warm-humid temperatures of the tropics do not induce reproductive cell differentiation,

| Treatments | Number of flowers | length of lateral branches (cm) | Number of leaves on lateral branches | Fruits number |
|---|---|---|---|---|
| Control (water) | 13.6 b* | 54.3 ab | 13.6 ab | 37.0 c |
| HBr (0.5%) | 19.0 ab | 60.6 ab | 15.4 a | 44.2 ab |
| HBr (1.0%) | 17.6 b | 65.9 a | 14.3 ab | 52.9 a |
| $KNO_3$ (1%) | 13.8 b | 54.1 ab | 12.4 ab | 29.0 c |
| HBr (0.5%) + $KNO_3$ (1%) | 27.1 a | 55.3 ab | 14.1 ab | 41.1 b |
| HBr (1%) + $KNO_3$ (1%) | 17.8 ab | 61.9 ab | 14.3 ab | 40.8 b |
| CV % | 31.0 | 22.0 | 21.7 | 20.5 |

**The letters that are not equal between the columns indicate significant statistical difference (Tukey $p \leq 0.05$). The application of the four treatments was carried out in February: HBr Cidef-4 al 0.5%, HBr Cidef-4 al 1%, $KNO_3$ al 1%, HBr Cidef-4 al 0.5% + $KNO_3$ al 1%, HBr Cidef-4 al 1%, + $KNO_3$ al 1% + and Control (water application). The plantation of C. arabica L. variety Catimor with a shade of Inga micheliana Harms. In each tree, a branch was identified toward each cardinal point and in it, an area of 50 cm was indicated, for the taking of variables of flowering and fruiting.*

**Table 7.**
*Comparisons of means of the number of flowers, leaves, fruits and length of four bands in* Coffea arabica *L. var. Catimor in response to the application of HBr in interaction with potassium nitrate.*

has been increased by potassium nitrate. Potassium nitrate as a flowering inducer and brassinosteroidal steroidal hormone can favor fruit growth in *Coffea* spp., as in *Mangifera indica* L [87]. The application of HBr (Cidef-4) alone and in combination with 1% $KNO_3$ was evaluated.

The number of flowers increased with the application of potassium nitrate plus HBr in the lowest concentration and was higher than the rest of the treatments ($P \leq 0.05$). It represented an increase of 100% in relation to the control. The lowest number of flowers was recorded in the control and when only potassium nitrate was applied. The HBr applications registered a mean increase in the number of flowers when it was applied alone, or in interaction with potassium nitrate, however, the highest values were presented when the concentration of 0.5% of HBr was included (**Table 7**).

The application of 1% KNO3 alone induced flowering similar to the control. On the other hand, with the application of the two concentrations of brassinosteroid alone, the difference in the number of flowers was 36% in relation to the control.

The growth of the bandola and the number of leaves increased in the treatments where only HBr was applied at the two concentrations, or in combination with potassium nitrate and the lowest values were with the application of potassium nitrate alone and the control.

The application of HBr induces a wide range of responses, including an increase in the cellular expansion of the leaves, increased elongation of the stem [14, 15, 28, 88] that increase the leaf surface, plant biomass and the yield of various crops [8].

The average number of fruits increased 29% more per band with the 0.5 and 1% HBr treatments. The combination of 1% KNO3 plus 0.5 and 1% of HBr induced similar values to the applications of HBr alone.

In different crops of economic importance, brassinosteroids are characterized by stimulating plant growth, increasing the yield of biomass production and accelerating the ripening of fruits. In addition, they strengthen the resistance of plants to pests and abiotic stressors such as drought and sudden changes in temperature [63].

It was discovered that Br promotes tomato yield through improved autophosphorylation of SlBRI1, and increased plant expansion, leaf area, fruit weight and number of fruit per cluster [10]. Brassinolide (BL), 28-homobrassinolide (28-hBL) and 24-epi-brassinolide (24-eBL) treatments stimulated both tomato growth and yield [41].

## 3. Conclusions

HBr favors the growth of *Musa* spp. *in vitro* through regrowth height and similarly to BAP and AIA regulators in the number of leaves and roots. In *ex vitro* conditions, there were differences in the growth of *Musa* spp. and *S. officinarum* between the application frequencies and in general, the increase in the number of applications favors this process.

The exogenous applications of HBr in *T. cacao* L, *M. indica* L. and *C. arabica* L. show variable results. Plant growth is modified as in *C. arabica* L and flowering is influenced as in *T. cacao* L and *C. arabica* L. and the yield in both crops. The higher concentrations applied did not necessarily increase the crop yield or the combination with potassium nitrate. The quality of the mango fruits showed variations in the concentrations of HBR but it was better alone or in combination with nitrate. Under field conditions, we consider it important to identify the dose and timing of HBr application in the plant organ of interest, be it flowers, foliage or fruits.

## Author details

Juan F. Aguirre-Medina[1], Jorge Cadena-Iñiguez[2] and Juan F. Aguirre-Cadena[1*]

1 Autonomous University of Chiapas, Tapachula, Chiapas, Mexico

2 Postgraduate College, Texcoco, Mexico State, Mexico

*Address all correspondence to: juan.cadena@unach.mx

## References

[1] Mitchell JW, Mandava NB, Worley JF, Plimmer JR, Smith MV. Brassins: A new family of plant hormones from rape pollen. Nature. 1970;**225**(5237):1065-1066. DOI: 10.1038/2251065a0

[2] Nuñez MP, Mazzafera LM, Mazorra W, Siqueira J, Zullo MA. Influence of brassinosteroid analogue on antioxidant enzymes in rice grown in culuture medium with NaCl. Biologia Plantarum. 2003;**47**:67-70. DOI: 10.1023/A:1027380831429

[3] Sharma P, Bhardwaj R. Effects of 24-epibrassinolide on growth and metal uptake in *Brassica juncea* L. under copper metal stress. Acta Physiologiae Plantarum. 2007;**29**(3):259-263. DOI: 10.1007/s11738-007-0032-7

[4] Bhardwaj R, Arora N, Sharma P, Arora HK. Effects of 28-homobrassinolide on seedling growth, lipid peroxidation and antioxidative enzyme activities under nickel stress in seedlings of *Zea mays* L. Asian Journal of Plant Sciences. 2007;**6**(5):765-772. DOI: 10.3923/ajps.2007.765.772

[5] Ahammed GJ, Li X, Liu A, Chen S. Brassinosteroids in plant tolerance to abiotic stress. Journal of Plant Growth Regulation. 2020;**39**:1451-1464. DOI: 10.1007/s00344-020-10098-0

[6] Sasse JM. Physiological actions of brassinosteroids: An update. Journal of Plant Growth Regulation. 2003;**22**:276-288. DOI: 10.1007/s00344-003-0062-3

[7] Li J, Jin H. Regulation of brassinosteroid signalling. Trends in Plant Sciences. 2007;**12**:37-41. DOI: 10.1016/j.tplants.2006.11.002

[8] Núñez M, Mazorra LM. Los brasinoesteroides y la respuesta de las plantas al estrés. Cultivos Tropicales. 2001;**22**(3):19-26 Available at: http://scielo.sld.cu/scielo.php?script=sci_arttext&pid=S0258-59362010000200008

[9] Smith SM, Li C, Li J. Hormone function in plants. In: Li J, Li C, Smith SM, editors. Hormone Metabolism and Signaling in Plants. London, United Kingdom: Elsevier; 2017. pp. 1-38. DOI: 10.1016/B978-0-12-811562-6.00001-3

[10] Wang S, Liu J, Zhao T, Du C, Nie S, Zhang Y, et al. Modification of Threonine-1050 of SlBRI1 regulates BR signalling and increases fruit yield of tomato. BMC Plant Biology. 2019;**19**(1): 256. DOI: 10.1186/s12870-019-1869-9

[11] Krishna P. Brassinosteroid-mediated stress response. Journal of Plant Growth Regulation. 2003;**22**(4):289-297. DOI: 10.1007/s00344-003-0058-z

[12] Yin H, Chen Q, Yi M. Effects of short-term heat stress on oxidative damage and responses of antioxidant system in *Lilium longiflorum*. Plant Growth Regulation. 2008;**54**(1):45-54. DOI: 10.1007/s10725-007-9227-6

[13] Li X, Zhang L, Ahammed GJ, Li ZX, Wei JP, Shen C, et al. Nitric oxide mediates brassinosteroid-induced flavonoid biosynthesis in *Camellia sinensis* L. Journal of Plant Physiology. 2017;**214**:145-151. DOI: 10.1016/j.jplph.2017.04.005

[14] Singh I, Shono M. Physiological and molecular efects of 24-epibrassinolide, a brassinosteroid on thermotolerance of tomato. Journal of Plant Growth Regulation. 2005;**47**(3):111-119. DOI: 10.1007/s10725-005-3252-0

[15] Reyes Y, Mazorra LM, Núñez M. Aspectos fisiológicos y bioquímicos de la tolerancia del arroz al estrés salino y su relación con los brasinoesteroides. Cultivos Tropicales. 2008;**29**(4):67-75 Available at: https://www.redalyc.org/pdf/1932/193214911012.pdf

[16] Bajguz A, Hayat S. Effects of brassinosteroids on the plant responses to environmental stresses. Plant Physiology and Biochemistry. 2009;**47**:1-8. DOI: 10.1016/j.plaphy.2008.10.002

[17] Vriet C, Russinova E, Reuzeau C. Boosting crop yields with plant steroids. The Plant Cell. 2012;**24**:842-857. DOI: 10.1105/tpc.111.094912

[18] González OJL, Córdova A, Aragón CE, Pina D, Rivas M, Rodríguez R. Effect of an analogue of brassinosteroid on FHIA-18 plantlets exposed to thermal stress. InfoMusa. 2005;**14**(1):18-20 Available at: https://eurekamag.com/research/004/118/004118289.php

[19] Núñez VM, Reyes GY, Rosabal AL, Martínez GL. Análogos espirostánicos de brasinoesteroides y sus potencialidades de uso en la agricultura. Cultivos Tropicales. 2014;**35**(2):34-42 Available at: http://scielo.sld.cu/scielo.php?script=sci_arttext&pid=S0258-59362014000200005

[20] Nie S, Huang S, Wang S, Cheng D, Liu J, Lv S, et al. Enhancing brassinosteroid signaling via overexpression of tomato (*Solanum lycopersicum*) SlBRI1 improves major agronomic traits. Frontiers in Plant Science. 2017;**8**:1386. DOI: 10.3389/fpls.2017.01386

[21] Zhou Y, Xia X, Yu G, Wang J, Wu J, Wang M, et al. Brassinosteroids play a critical role in the regulation of pesticide metabolism in crop plants. Scientific Reports. 2015;**5**:9018. DOI: 10.1038/srep09018

[22] Serna M, Coll Y, Zapata PJ, Botella MA, Pretel MT, Amorós A. A brassinosteroid analogue prevented the effect of salt stress on ethylene synthesis and polyamines in lettuce plants. Scientia Horticulturae. 2015;**185**:105-112. DOI: 10.1016/j.scienta.2015.01.005

[23] Fariduddin Q, Khanam S, Hasan SA, Ali B, Hayat SA, Ahmad A. Effect of 28-homobrassinolide on the drought stress-induced changes in photosynthesis and antioxidant system of *Brassica juncea* L. Acta Physiologiae Plantarum. 2009;**31**:889-897. DOI: 10.1007/s11738-009-0302-7

[24] Hasan M, Ali Md. A, Soliman MH, Alqarawi AA, Abd Allah EF, Fang XW. Insights into 28-homobrassinolide (HBR)- mediated redox homeostasis, AsA–GSH cycle, and methylglyoxal detoxification in soybean under drought-induced oxidative stress. Journal of Plant Interactions. 2020; 15(1): 371-385. DOI: 10.1080/17429145.2020.1832267

[25] Esparza M. La sequía y la escasez de agua en México. Situación actual y perspectivas futuras. Secuencia. 2014;**89**:195-219 Available at: http://www.scielo.org.mx/pdf/secu/n89/n89a8.pdf

[26] Zullo MAT, Kohout L, De Azevedo MBM. Some notes on the terminology of brassinosteroids. Journal of Plant Growth Regulations. 2003;**39**:1-11. DOI: 10.1023/A:1021802910454

[27] Rao SSR, Vardhini BV, Sujatha E, Anuradha S. Brassinosteroids – A new class of phytohormones. Current Science. 2002;**82**(10):1239-1245 Available at: https://www.jstor.org/stable/24107046

[28] Nemhauser JL, Chory J. Bring it on new insights into the mechanism of brassinosteroid action. Journal of Experimental Botany. 2004;**55**(395): 265-270. DOI: 10.1093/jxb/erh024

[29] Chai YM, Ch-Li L, Zhang Q, Xing Y, Tian L, Qin L, et al. Brassinosteroid is involved in strawberry fruit ripening. Journal of Plant Growth Regulation. 2013;**69**(1):63-69. DOI: DOI. 10.1007/s10725-012-9747-6

[30] Clouse SD. Recent advances in brassinosteroid research: From molecular mechanisms to practical applications. Journal of Plant Growth Regulation. 2003;**22**:273-275. DOI: 10.1007/s00344-003-0067-y

[31] Müssig C. Brassinosteroid-promoted growth. Plant Biology. 2005;**7**(2):110-117. DOI: 10.1055/s-2005-837493

[32] Salgado GR, Cortés RMA, Del Río RE. Uso de brasinoesteroides y sus análogos en la agricultura. Biológicas. 2008;**10**:18-27 Available at: https://exa.unne.edu.ar/biologia/fisiologia.vegetal/Brasinoesteroidesyan%C3%A1logosenlaagricultura.pdf

[33] Izquierdo H, González MC, Núñez M, Proenza R, Álvarez I. Efectos de la aplicación de un análogo espirostánico de brasinoesteroides en vitroplantas de banano (*Musa* spp.) durante la fase de aclimatización. Cultivos Tropicales. 2012;**33**(1):71-76. Available at: http://scielo.sld.cu/scielo.php?pid=S0258-59362012000100011&script=sci_abstract

[34] Bao-Fundora L, Hernández Ortiz RM, Diosdado Salcés E, Román Gutiérrez MI, González Arencibia C, Rojas Álvarez A, et al. Embriogénesis somática de Citrus macrophylla Wester con el empleo del Pectimorf® y análogos de brasinoesteroides. Revista Colombiana de Biotecnología. 2013;**XV**(1):192-172 Available at: https://www.redalyc.org/pdf/776/77628609021.pdf

[35] Sirhindi G, Kumar M, Bhardwaj R, Kumar S, Pradhan SK. Effect of 24-epibrassinolide on activity of antioxidant enzymes in *Brassica juncea* L. under $H_2O_2$ stress. Indian Journal of Plant Physiology. 2011;**16**(1):68-71. DOI: 10.1007/s12298-009-0038-2

[36] Vleesschauwer DD, Buyten EV, Satoh K, Balidion J, Mauleon R, Choi IR, et al. Brassinosteroids antagonize gibberellin and salicylate-mediated root immunity in rice. Plant Physiology. 2012;**158**:1833-1846. DOI: 10.1104/pp.112.193672

[37] De León RJM, Gálvez López AL, Aguirre-Medina JF. Aplicación de brasinoesteroide a seis orquídeas para crecimiento in vitro. In: Cih-Dzul IR, Moreno HA, FFJ C, BVM S, MCE G, editors. Alimentación sostenible y retos del sistema agroalimentario. Primera Edición. Zapopan Jalisco, México: Editorial Página Seis, S.A. de C.V; 2018. pp. 517-536

[38] Murashige T, Skoog F. A revised medium for rapid growth and bioassays with tobacco tissue cultures. Physiologia Plantarum. 1962;**15**:437-497. DOI: 10.1111/j.1399-3054.1962.tb08052.x

[39] Aguirre-Medina JF, Gálvez López AL, De León RJM, Rodas MN. Explantes para la obtención de callos embriogénicos somáticos en Stevia rebaudiana (Bert.). In: Cih-Dzul IR, Moreno HA, Cárdenas FFJ, BVM S, MCE G, editors. Alimentación sostenible y retos del sistema agroalimentario. Primera Edición. Zapopan Jalisco, México: Editorial Página Seis, S.A. de C.V; 2018. pp. 457-476

[40] Aguirre Medina JF, Mina Briones FO, Cadena Iñiguez J, Soto Hernández RM. Efectividad de biofertilizantes y brasinoesteroide en *Stevia rebaudiana* Bert. Agrociencia. 2018;**52**(4):609-621 Available at: https://www.colpos.mx/agrocien/Bimestral/2018/may-jun/art-10.pdf

[41] Vardhini BJ, Rao SSR. Effect of brassinosteroids on growth and yield of tomato (*Lycopersicon esculentum* mill.) under field conditions. Indian Journal of Plant Physiology. 2001;**6**(3):326-328

[42] Fu FQ, Mao WH, Shi K, Zhou YH, Asami T, Yu JQ. A role of brassinosteroids in early fruit development in cucumber. Journal of

Experimental Botany. 2008;**59**(9):2299-2308. DOI: 10.1093/jxb/ern093

[43] Hayat S, Ahmad A, Mobin M, Fariduddin Q, Azman ZM. Carbonic anhydrase, photosynthesis, and seed yield in mustard plants treated with Phytohormones. Photosynthetica. 2001;**39**:111-114. DOI: 10.1023/A:1012456205819

[44] Janeczko A, Biesaga-Koscielniak J, Oklestkova J, Filek M, Dziurka M, Szarek-Łukaszewska G, et al. Role of 24-epibrassinolide in wheat production: Physiological effects and uptake. Journal of Agronomy and Crop Science. 2010;**196**:311-321. DOI: 10.1111/j.1439-037X.2009.00413.x

[45] Hayat S, Maheshwari P, Wani AS, Irfan M, Alyemeni MN, Ahmad A. Comparative effect of 28 homobrassinolide and salicylic acid in the amelioration of NaCl stress in *Brassica juncea* L. Plant Physiology and Biochemistry. 2012;**53**(1):61-68. DOI: 10.1016/j.plaphy.2012.01.011

[46] Peng J, Tang X, Feng H. Effects of brassinolide on the physiological properties of litchi pericarp (*Litchi chinensis* cv. Nuomoci). Scientia Horticulturae. 2004;**101**:407-416. DOI: 10.1016/j.scienta.2003.11.012

[47] Azofeifa A. Problemas de oxidación y oscurecimiento de explantes cultivados in vitro. Agronomía Mesoamericana. 2009;**20**(1):153-175 Available at: http://www.mag.go.cr/rev_meso/v20n01_153.pdf

[48] Aragón C, Carvalho L, González J, Escalona M, Amancio S. *Ex vitro* acclimatization of plantain plantlets micropropagated in temporary immersion bioreactor. Biologia Plantarum. 2010;**54**(2):237-244. DOI: 10.1007/s10535-010-0042-y

[49] Ali B, Hayat S, Fariduddin Q, Ahma A. 24-Epibrassinolide protects against the stress generated by salinity and nickel in *Brassica juncea*. Chemosphere. 2008;**72**(9):1387-1392. DOI: 10.1016/j.chemosphere.2008.04.012

[50] Jiménez FA, Ramírez D, Agramonte D. Use of Biobras-6 in micropropagation of 'FHIA-21'. InfoMusa. 2004;**13**(1):4-6 Available at: https://www.bioversityinternational.org/fileadmin/_migrated/uploads/tx_news/Infomusa__La_revista_internacional_sobre_bananos_y_pl%c3%a1tanos_966.pdf

[51] Ferdous MH, Masum BAA, Mejraj H, Taufique T, Jamal UAFM. BAP and IBA pulsing for in vitro multiplication of banana cultivars through shoot-tip culture. Journal of bioscience and agriculture. Research. 2015;**03**(02):87-95. Available at: www.journalbinet.com/jbar-journal.html

[52] Ali MR, Mehraj H, Uddin JAFM. Kinetin (KIN) and Indole-3-acetic acid (IAA) on *in vitro* shoot and root initiation of tuberose. International Journal Sustainable Agricultural Technology. 2014;**10**(8):1-4 Available at SSRN: https://ssrn.com/abstract=3599034

[53] Madhulatha P, Anbalagan M, Jayachandran S, Sakthivel N. Influence of liquid pulse treatment with growth regulators on *in vitro* propagation of Banana (*Musa* spp. AAA). Plant Cell, Tissue and Organ Culture. 2004;**76**:189-192. DOI: 10.1023/B:TICU.0000007291.31439.6c

[54] Buah JN, Danso E, Taah KJ, Abole EA, Bediako EA, Asiedu J, et al. The effects of different concentrations cytokinins on the *in vitro* multiplication of plantain (*Musa* sp). Biotechnology. 2010;**9**(3):343-347 Available at: https://scialert.net/abstract/?doi=biotech.2010.343.347

[55] North J, Ndakidemi P, Laubscher C. Effects of antioxidants, plant growth

regulators and wounding on phenolic compound excretion during micropropagation of *Strelitzia Reginae*. International Journal of Physical Sciences. 2012;**7**(4):638-646. DOI: 10.5897/IJPS11.786

[56] Hernández RM, Moré O, Núñez M. Empleo de análogos de brasinoesteroides en el cultivo *in vitro* de papa (*Solanum tuberosum* L.). Cultivos Tropicales. 1999;**20**(4):41-44 https://www.redalyc.org/pdf/1932/193230162005.pdf

[57] Nassar AH. Effect of homobrassinolide on *in vitro* growth of apical meristems and heat tolerance of banana shoots. International Journal of Agriculture & Biology. 2004;**6**(5):771-775 Available at: https://eurekamag.com/research/004/122/004122682.php

[58] Robres-Torres E, López-Medina J, Rocha-Granados MC. La elongación de brotes adventicios de frambuesa (*Rubus ideaus* L.) es influenciada por brasinosteroides. Revista Mexicana de Ciencias Agrícolas. 2015;**6**(5):991-999. Available at: https://www.redalyc.org/pdf/2631/263139893006.pdf

[59] Terry AE, Ruiz PJ, Tejeda PT, Dra SC, Reynaldo EI, Díaz de Armas MM. Respuesta del cultivo de la lechuga (*Lactuca sativa* L.) a la aplicación de diferentes productos bioactivos. Cultivos Tropicales. 2011;**32**(1):28-37. Available at: http://scielo.sld.cu/scielo.php?pid=S0258-59362011000100003&script=sci_arttext&tlng=pt

[60] Cortes PA, Terrazas T, Colinas León T, Larque-Saavedra A. Brassinosteroid effects on precocity and yield of cladodes of cactus pear (*Opuntia ficus-indica* L. mill). Scientia Horticulturae. 2003;**97**:65-73. DOI: 10.1016/S0304-4238(02)00080-8

[61] Suárez L. Efecto que ejercen las aspersiones foliares de una mezcla de oligogalacturónidos (Pectimorf) y la formulación a base de un análogo de brasinoesteroides (Biobras-16) en dos especies de orquídeas (*Cattleya leuddemanniana* y *Guarianthe skinneri*). Cultivos Tropicales. 2007;**28**(4):87-91 Available at: https://www.redalyc.org/pdf/1932/193217894014.pdf

[62] Raut S, Ranade S. Diseases of Banana and their management. In: Naqvi SAMH, editor. Diseases of Fruits and Vegetables: Volume II. Dordrecht: Springer; 2004. pp. 37-52. DOI: 10.1007/1-4020-2607-2_2

[63] Mandava NB. Plant growth-promoting Brassinosteroids. Annual Review of Plant Physiology and Plant Molecular Biology. 1988;**39**(1):23-52. DOI: 10.1146/annurev.pp.39.060188.000323

[64] Jeyakumar P, Kumar N, Kavino M. Physiological response of banana cv. Robusta (AAA) to foliar applied plant growth regulators on productivity. Madras Agricultural Journal. 2003;**90**(10-12):702-706

[65] Herrera AJ, Aguirre-Medina JF, Gálvez AL, Ley de Coss A, Martínez M. Efecto de reguladores de crecimiento en la reproducción in vitro de *Musa* spp cv gran enano. AGROproductividad. 2017;**10**(9):20-25 Available at: http://www.colpos.mx/wb/index.php/agroproductividad#.WnoBa1TibIU

[66] Li Q-F, Lu J, Zhou Y, Wu F, Tong H-N, Wang J-D, et al. Abscisic acid represses rice lamina joint inclination by antagonizing brassinosteroid biosynthesis and signaling. International Journal of Molecular Sciences. 2019;**20**(19):4908. DOI: 10.3390/ijms20194908

[67] Capote I, Escalona M, Gradaille MD, Pina D, González JL, Aragón C. Efecto del análogo de Brasinoesteroide (MH5) en la aclimatación de los brotes de *Vriesea* propagados en sitemas de inmersión temporal. Revista Ciencia y

Tecnología. 2009;**2**(1):29-33 Available at: https://dialnet.unirioja.es/servlet/articulo?codigo=4054769

[68] Chavana CS, Chavana SS, Velhala RS, Patila AG, Deshmukhb LP, Bapat VA. Effect of Brassinolide on Invitro growth of sugarcane Co 86032 (Saccharum spp.). In: AIP Conference Proceedings 1989. American Institute of Physics; 2018. p. 030004. Published by the. DOI: 10.1063/1.5047722

[69] Cao S, Xu Q, Cao Y, Qian K, An K, Zhu Y, et al. Loss-of-function mutations in DET2 gene lead to an enhanced resistance to oxidative stress in *Arabidopsis*. Physiologia Plantarum. 2005;**123**(1):57-66. DOI: 10.1111/j.1399-3054.2004.00432.x

[70] Montoya T, Nomura T, Yokota T, Farrar K, Harrison K, Jones JGD, et al. Patterns of dwarf expression and brassinosteroid accumulation in tomato reveal the importance of brassinosteroid synthesis during fruit development. The Plant Journal. 2005;**42**(2):262-269. DOI: 10.1111/j.1365-313X.2005.02376.x

[71] Estrada-Lugo EIJ. El códice Florentino, su información etnobotánica. Chapingo, Edo. De México. México: Ed. Colegio de Posgraduados; 1989. p. 399

[72] Ogata N. El cacao. Biodiversitas. 2007;**72**:1-5 Available at: http://infocafes.com/portal/wp-content/uploads/2016/01/biodiv72art1.pdf

[73] García OJ, Yiovani Duéñez E, Fischer G, Chaves B, Quintero OC. Fruit set response of pineapple guava (*Acca sellowiana* [O. berg] Burret) to potassium nitrate, potassium phosphate and ethephon. Agronomía Colombiana. 2008;**26**(2):217-225. Available at: https://www.researchgate.net/publication/256474809_El_cuajamiento_de_frutos_de_feijoa_Acca_sellowiana_O_Berg_Burret_en_respuesta_a_nitrato_de_potasio_fosfato_de_potasio_y_ethephon

[74] Hayat S, Ahamad A, Mobin M, Hussain A, Fariduddin Q. Photosynthetic rate, growth and yield of mustard plants sprayed with 28-homobrassinolide. Photosynthetica. 2000;**38**(3):469-471. DOI: 10.1023/A:1010954411302

[75] Gomes M, Lagôa A, Machado E, Medina CL. Abscisic acid and indole-3-acetic acid contents in orange trees infected by Xylella fastidiosa and submitted to cycles of water stress. Plant Growth Regulation. 2003;**39**:263-270. DOI: 10.1023/A:1022854405898

[76] Zullo TMA, Adam G. Brassinosteroid phytohormones - structure, bioactivity and applications. Brazilian Journal of Plant Physiology. 2002;**14**(3):143-181 Available at: https://www.scielo.br/j/bjpp/a/jt9wVShx5fgHCFbxhBhRFPn/?lang=en&format=pdf

[77] Gomes MMA, Campostrini E, Rocha Leal N, Pio Viana A, Massi Ferraz T, Siqueira LN, et al. Brassinosteroid analogue effects on the yield of yellow passion fruit plants (*Passiflora edulis* f. flavicarpa). Scientia Horticulturae. 2006;**110**(3):235-240. DOI: 10.1016/j.scienta.2006.06.030

[78] Ali B, Hayat S, Ahmad A. 28-Homobrassinolide ameliorates the saline stress in chickpea (*Cicer arietinum* L.). Environmental and Experimental Botany. 2007;**59**(2):217-223. DOI: 10.1016/j.envexpbot.2005.12.002

[79] Mazorra LM, Oliveira MG, Souza AF, Da silva WB, Dos Santos GM, Almeida Da silva LR, Gomez da Silva M, Bartoli CG, De Oliverira GJ. Involvement of brassinosteroids and ethylene in the control of mitochondrial electron transport chain in postharvest papaya fruit. Theoretical and experimental Plant Physiology 2013; 25(3): 223-230. Available at: https://www.scielo.br/j/txpp/a/WPfhpZKG43X7cnpszJsh7Nw/?format=pdf&lang=en

[80] Susila T, Reddy SA, Rajkumar M, Padmaja G, Rao P. Effects of sowing date and spraying of brassinosteroid on yield and fruit quality characters of watermelon. World Journal of Agricultural Sciences. 2012;**8**(3):223-228 Available at: https://www.idosi.org/wjas/wjas8(3)12/1.pdf

[81] De Grauwe L, Vandenbussche F, Tietz O, Palme K, Van Der Straeten D. Auxin, ethylene and brassinosteroids: Tripartite control of growth in the *Arabidopsis* hypocotyl. Plant and Cell Physiology. 2005;**46**(6):827-836. DOI: 10.1093/pcp/pci111

[82] Arteca RN, Arteca JM. Effects of brassinosteroid, auxin, and cytokinin on ethylene production in *Arabidopsis thaliana* plants. Journal of Experimental Botany. 2008;**59**(11):3019-3026. DOI: 10.1093/jxb/ern159

[83] Zaharah SS, Singh Z, Symon GM, Reid JB. Role of brassinosteroids, ethylene, abscisic acid, and indole-3-acetic acid in mango fruit ripening. Journal of Plant Growth Regulations. 2012;**31**:363-372. DOI: 10.1007/s00344-011-9245-5

[84] Symons GM, Chua YJ, Ross JJ, Quittenden LJ, Davies NW, Reid JB. Hormonal changes during non-climacteric ripening in strawberry. Journal of Experimental Botany. 2012;**63**:4741-4750. DOI: 10.1093/jxb/ers147

[85] Azcón-Bieto J, Talón M. Fundamentos de Fisiología Vegetal. Madrid España: Editorial Interamericana McGraw-Hill; 2008. p. 669

[86] Davenport T. Reproductive physiology of mango. Brazilian Journal of Plant Physiology. 2007;**19**(4):363-376 Available at: https://www.scielo.br/j/bjpp/a/M3wyHvkcRMyrBrkTrzvzjyz/?format=pdf&lang=en

[87] Aguirre-Medina JF, Hernández-quívez A, Arévalo-Galarza L, Ley De Coss A, Gehrke-Velez MR. Application of a steroidal hormone affects fruit yield and quality in 'Ataúlfo' mango (*Mangifera indica* L.). Indian Journal of Horticulture. 2015;**5**(3/4):86-92. Available at: https://www.indianjournals.com/ijor.aspx?target=ijor:ihj1&volume=5&issue=3and4&article=006

[88] Hu Y, Bao F, Li J. Promotive effect of brassinosteroids on cell division involves a distinct CycD3-induction pathway in Arabidopsis. Plant Journal. 2000;**24**(5):693-701. DOI: 10.1046/j.1365-313x.2000.00915.x

Chapter 10

# Genetic Modification of Plant Hormones Induced by Parasitic Nematodes, Virus, Viroid, Bacteria, and Phytoplasma in Plant Growing

*Refik Bozbuga, Bekir Bulent Arpaci, Selman Uluisik, Pakize Gok Guler, Hatice Nilufer Yildiz and Songul Yalcin Ates*

## Abstract

Plant hormones, such as auxin, play crucial roles in many plant developmental processes, which is crosstalk with gibberellin and strigolactone. The roles of hormones may vary in the biosynthesis of metabolisms. During the pathogen attack, including plant-parasitic nematodes, viroid, phytoplasma, virus, and bacteria, plant hormones are involved in several plant processes. Ethylene (ET), salicylate (SA), jasmonate (JA), and abscisic acid (ABA) primarily regulate synergistically or antagonistically against pathogens. Those pathogens—nematodes, bacteria, viroid, phytoplasma, and viruses regulate several plant hormones for successful parasitism, influencing the phytohormone structure and modifying plant development. Several genes are related to plant hormones that are involved in pathogens parasitism. In this chapter, how pathogens affect plant hormones in plants growing are discussed.

**Keywords:** plant hormones, gene, parasitic nematodes, bacteria, virus, phytoplasma, viroid, phytoplasma

## 1. Introduction

Plant hormones, small chemicals, also termed phytohormones, play an amazing array of very important roles in plant development, fruit ripening, from embryogenesis to senescence and stress resistance. Among the nine well-characterized plant hormones, gibberellin (GA), strigolactones (SLs), brassinosteroids (BRs), cytokinins (CKs), and auxin are classified as growth-promoting hormones, and others, including jasmonic acid (JA), salicylic acid (SA), ethylene, and abscisic acid (ABA) are considered stress reaction hormones. Several studies reveal that the role of hormones in the regulation of several physiological processes in diverse tissues or dissimilar ecological conditions has widely been described [1, 2]. Although little is known about the interaction between hormones at various levels of these cascades nevertheless, with the initiation of "omics" tools, noteworthy development has been made in the description of phytohormone reactions in different stimuli [3].

Although there are many studies in which exogenously hormone treatment is made to increase the plant quality [4], plant hormones are generated by the plants themselves, and their functions and chemical structures are mainly conserved among plants. Since these combinations are signal molecules, their cellular intensities are firmly controlled at the level of catabolism, biosynthesis, and transport in reaction to growing and ecological conditions. Therefore, here we have surveyed the published literature on the role of phytohormones in the perspective of their effects on numerous features of plant growth and respond to biotic stress conditions, such as plant-parasitic nematodes, virus, viroid, bacteria, and phytoplasma, and describe ways in which phytohormones modulate these reactions separately or grouping with several hormones. In addition to biotic stress and hormone interactions, special attention is also paid to the tomato, a well-studied plant, for the molecular genetic investigation of succulent plant growth and fruit ripening.

## 2. The role of phytohormones in plant development fruits ripening

The hormone auxin plays a key role in organ development as well as in distinct aspects of plant growth and development, such as cell elongation, division, and differentiation. Among auxin hormones, indole-3-acetic acid (IAA) is the most abundant occurring form in plants. Leaf primordia are known as the major sites for the synthesis of auxins, but roots contain much higher levels of auxins due to the polar transport of this hormone [5]. Construction of adventitious root (AR) is an important developmental process in cutting propagation for the horticultural industry. AR formation is divided into three developmental stages—induction, initiation, and lengthening [6, 7], and it has been stated that auxin (IAA) regulates the formation of AR at almost every developmental stage of these three stages. Although Arabidopsis thaliana have been well characterized in terms of mechanism of AR formation, there is no detailed study in other species involving tomato in which stem cuttings readily form ARs that which them an ideal system to study AR formation in detail. It was shown that AR is formed in tomato stem cuttings after the perception of a wounding stimulus [8]. The effect of auxin on the transcriptional regulation of target genes, through Auxin Response Factors (ARFs), has been shown to have the potential involvement of many ARFs in a/biotic stress responses [9]. For example, overexpression of an auxin-responsive gene, TaSAUR75, in Arabidopsis altered root length and survival rate and higher expression of some stress-sensitive genes, increasing drought and salt tolerance [10]. ARF genes also govern the initiation of fruit set by the involvement of SlARF8, SlARF7, and SlARF9 on plant fruit development through the interactions between the auxin- and GA-signaling pathways [11, 12]. During tomato fruit ripening, auxin induces ethylene biosynthesis through upregulation of both *ACS* and *ACO* genes, and *Sl-SAUR69*, auxin-related gene, plays an important role in the initiation of "system 2" ripening, where the fruit experiences high sensitivity to ethylene [13]. Abscisic acid (ABA) is one of the most important hormones in the response mechanism against stress conditions and plays an important role in germination, root development, drought tolerance, and growth [14]. It is the best-known stress signaling molecule in plants, and it defends plants from a/biotic stresses. Apart from this, ABA controls almost the entire process of fruit ripening by regulating the ABA-ethylene pathways [15]. Conserved pyrabactin resistance/pyrabactin resistance-like/regulatory component of ABA receptors (PYR/PYL/RCAR) sense the ABA and trigger a series of signaling events. The hormone is normally known as plant growth inhibitor. However, the experiments showed that ABA promotes growth on roots and shoots. For instance, exogenous treatment of ABA to plant shoots encourage primary root (PR) growth [16]. The main function of ABA in plants reported that it has a key role in stress adaptation involved in drought and salinity stress. It has been shown that ABA accumulation increases up to 40 times in plants exposed to drought stress. Drought and salinity stresses activate ABA biosynthetic genes through activating ABA response genes by binding ABA-responsive elements (ABREs) located in their promoter regions and increasing ABA levels in plants [17]. Mutant plants that are insensitive to ABA have also been identified as sugar insensitive, indicating a strong correlation between ABA and sugar signaling [18]. It was recently shown that increased ABA hormone content mediates the expression of drought and salt stress-induced *OsSWEET13* and *OsSWEET15* genes, resulting in increased sucrose content in phloem and altered sucrose transporting in leaf and root tissues [19].

ABA also plays a key role in plant immunity, but this involvement is more complex than its involvement in abiotic stress responses because the effect of ABA on the resistance of plants to fungi appears to depend on several factors, such as the rate of increase of the hormone, the type of pathogen, source of ABA (endogenous or exogenous), or application method. ABA-mediated response against biotic stresses is not fully understood. However, ABA also affects the expression of genes encoding proteins associated with the plant's response to a/biotic stresses and the spread of pathogens in the plant [20]. The functions of ABA in the pathogen defense mechanism may generally be related to SA, JA, and ET, whose signals are in more pathogen defense than ABA signaling [21]. In other words, endogenous ABA and JA form a synergy and exhibit a complex antagonistic relationship with SA during the development of the pathogen [22]. Based on the current studies, the research on ABA has become increasingly detailed, covering all aspects of plant development, fruit ripening, and a/biotic stress conditions, from properties and physiological indicators to the molecular and cellular levels, with greater emphasis on the study of various mechanisms.

Ethylene is the simplest and first molecule discovered as a hormone in plants [23]. Ethylene is synthesized by plants and influences various developmental processes, such as seed germination, fruit ripening, senescence, as well as responses to various a/biotic stresses. The ethylene signal transduction pathway has been broadly investigated in tomatoes because ethylene affects many characteristics related to plant vigor, fruit ripening, and postharvest physiology and storage. Numerous transcription factors have been identified that participate in the fruit ripening mechanism by regulating ethylene responses. The role of ethylene in the regulation of fruit ripening has been extensively reviewed [24]. Besides plant growth and development, recent studies have emphasized the role of ethylene in regulating plant responses to various stress conditions [25]. Ethylene is one of the important positive mediators for stress tolerance against salinity in many plants, such as the *Arabidopsis* model plant, maize, and tomato [26, 27]. Ethylene enhanced seed germination in saline conditions by reducing the expression of *MsACO* and *MsERF8* genes in alfalfa, showing that ethylene improves salt tolerance in alfalfa via *MsETR2* dependent manner [28]. In another recent study, the application of 1-MCP (an ethylene precursor) and AVG (an ethylene biosynthesis inhibitor), respectively, increased and decreased the cold resistance of apple seedlings. This study suggested that *MdERF1B*, an ethylene signaling activator, is a potential regulatory module that integrates the cold and ethylene signaling pathways in apples [29]. It is clear that ethylene plays a very important role in plant adaptation under abiotic stress. However, it would be more useful to determine the molecular cocktail of the antagonistic and synergistic role of ethylene with other signaling cues.

In addition to abiotic stress conditions, the role of ethylene in plants exposed to biotic stress has been investigated in different plant species and different diseases. Molecular studies in plants have shown in plants that many plant hormone-regulated pathways related to resistance to biotic stresses are stimulated by the pathogen through the biosynthesis of effector proteins [30]. However, the role of ethylene in host-defense mechanisms against pathogens is complex and could be controversial due to the conflicting roles that ethylene can increase susceptibility or resistance depending on the host-pathogen interaction studied [31]. However, recent evidence demonstrates that the ethylene signaling pathway is involved in the plant defense response against biotrophic and necrotrophic pathogens. In apple, over-expression of the *MdERF11* gene in apple callus significantly enhances resistance to *B. dothidea* infection, whereas down-regulation of the gene in apple callus reduced the resistance [28]. In a similar study, over-expression of *AcERF2*, an ethylene-responsive factor, increased the resistance against both fungal pathogen *B. cinerea* and bacterial pathogen Pst DC 3000 in transgenic plants by regulating the expression of biotic stress-responsive genes positively [32]. These results indicate that ethylene not only regulates several aspects of plant growth and fruit ripening but also participates in defense mechanisms against biotic stress factors.

Gibberellins (GAs) are another important hormone that controls plant growth by regulating various physiological mechanisms [33]. Like many other plant hormones, GAs play key roles in stem and root elongation, flowering, leaf enlargement, fruit senescence, seed germination, and increased fruit size. Exogenously applied studies have shown that GAs affect various physiological activities, such as vegetative growth, flowering fruit set, leaf area expansion, internode elongation, and can also increase biomass production, fruit weight, and dryness of plant organs [34, 35]. Apart from the roles associated with the growth and development of the plant, GAs plays a critical role in defense against environmental stresses. Treatment with GAs has been reported to increase the resistance to salinity stress. The foliar application of GA3 successfully elevated salinity tolerance of tomato and sweet pepper seedlings up to 25 mM NaCl and 50 mM NaCl, respectively showing the positive effect of GA3 in plants treated with salinity stress [36]. Exogenous GA increased the survival rate of wheat seedlings by regulating antioxidant defense mechanisms and the glyoxalase system under drought stress [37]. Moreover, treatment of tomato fruits with GA3 effectively reduced chilling injury (CI) index in mature green tomato fruit during long-term cold storage by reducing the expression of the DELLA protein, which is known as a crucial GA signaling component and growth repressor [38]. These findings provide a comprehensive insight into the mechanisms by which GA mediates fruit tolerance against abiotic stress conditions.

Although GA has received less attention n in the elucidation of signaling components involved in defense responses to biotic stress conditions. However, recent evidence shows that GA signaling components play an important role in plant disease resistance and sensitivity. It has been shown that DELLA proteins, which function as repressors in GA signaling pathways, control plant immune responses [39]. It was reported that overexpression of four *MeDELLAs* identified in cassava increased the disease resistance against cassava bacterial blight. In contrast, virus-induced gene silencing (VIGS) of the genes reduced the resistance with lower transcript levels of defense-related genes [40]. In a recent study, exogenous GA3 treatment promoted the resistance of rice to brown planthopper (BPH) [41]. Accumulating evidence suggests that GA and signaling components play important roles in regulating defense responses against a variety of pathogens too.

## 3. Phytohormone stimulating by plant-parasitic nematodes

Nematodes are mostly microscopic organisms that are found in many places on the earth, from oceans to continents. They may be parasitic or free-living that plant-parasitic nematodes cause damage to crops in the world. Plant-parasitic nematodes infect many plants, including crops, vegetables, and fruit trees [42], and they are the most damaging group of plant pathogens [43]. Root-knot nematodes (*Meloidogyne* ss) are one of the most damaging nematode genera in the world, and they are sedentary endoparasitic nematodes. Root-knot nematodes modify cell wall molecular architecture in their feeding site termed giant cells in successful parasitism [44, 45]. Plants also respond to nematode as a resistance gene such as SlWRKY3 roles a resistance against *Meloidogyne javanica* activating plant hormone [46]. Root-knot nematodes form a feeding tube where the glands insert the stylet to release nematode secretions to employ plant resistance and construct a feeding place [47]. Cell wall thickness change induced by nematodes [48]. Nematodes damage plants by affecting the phytohormone structure and altering plant growth. Plant hormones—auxin and cytokinin involve in a vital role in establishing established nematode feeding sites [49]. Auxin, cytokinin, and ethylene involve in gall formation after root-knot nematode infection in plant roots and activates the formation [50–53]. Auxin plays a significant role in cell division and growth of plant roots [53] that it also plays a role in gall development and gall expansion in roots of A. thaliana after Meloidogyne incognita infection [54].

Phytohormones: auxins, cytokinins, gibberellins, abscisic acid, and peptide hormones involved in plant defense. But, signaling pathways, but their function in plant defense is not well understood [55], and they involve in gene expression of plant defense and assembly of defense molecules such as pathogenicity-like proteins (PR) [56], phytoalexins [57] phenylpropanoids [58]. During the nematode infection, the concentrations of plant hormones [59]. Auxin, ethylene, cytokinin, gibberellic acid, abscisic acid, nitric oxide, jasmonic acid, brassinosteroids, salicylic acid, and strigolactones are plant hormones [60]. It appears that plant growth hormones, particularly cytokinin and auxin, are central to the formation of pathogen/pest-infected plant galls [61]. Root-knot nematodes and cyst nematodes modify auxin homeostasis thru several approaches [62]. Nematodes can alter several plant hormones for successful parasitism. Each hormone may coordinately stimulate potato tuberization formation or suppresses [63].

Gibberellins have functions such as controlling growth, metabolic activity, reactive oxygen species (ROS) in reaction during abiotic and biotic stresses [64, 65]. Gibberellin, glycosyltransferases, and auxins involve in biosynthesis and inactivation pathways in the nematode resistance response and their involvement in jasmonate signaling and biotic stress response in soybean-root-knot nematode interactions [66]. Several proteins are involved in hormone perception and signaling cytokinin, gibberellic acid, auxin, ethylene, and jasmonate recognition [60, 67].

Several genes are dissimilarly expressed in reaction to nematodes associated with plant hormones [60]. Several hormones are involved in nematode feeding site formation, such as ethylene biosynthesis in feeding cell formation by cyst nematodes, cytokinin, and ethylene-mediated regulatory networks in feeding cell development [60]. Auxin or IAA (indole-3-acetic acid) is involved in several developmental processes in plants. Auxin-mediated regulatory networks in nematode feeding cells, polar auxin transport (pat) plays a role in feeding cell formation and variation of local auxin intensities by endoparasitic nematodes [60].

Stress-related genes are notable genes involved in glycosyltransferases, peroxidases, auxin-sensitive proteins, and gibberellin-regulated genes during the Root-knot nematode in soybean infection. In the resistance reaction, auxins gibberellin signal transduction and glycosyltransferases, revealing the key role of components of biosynthesis and deactivation pathways and their involvement in jasmonate signaling and redox homeostasis involve in responses to biotic stress [66].

Cytokinin signaling is activated at nematode feeding sites to respond against sugar beet cyst nematode infection in the Arabidopsis plant [68].

Plant-parasitic nematodes may regulate phytohormone pathways that auxin and cytokinin involve in the formation of root-knot nematode and cyst nematodes feeding site, and plant genes are involved in hormone pathways [69]. Jasmonic acid defense molecules and salicylic acid-related defense are involved in nematode-plant interactions [69]. Ethylene plays a role in plant susceptibility to root-knot nematode and cyst nematodes [69]. Arabidopsis REPRESSOR OF CYTOKININ DEFICIENCY 1 transports UDP-N-acetylglucosamine/UDP-N-acetylgalactosamine and regulates endoplasmic reticulum protein quality control and cytokinin [70]. A transcription factor, WRKY23, is regulated by auxins and prompted by the cyst nematode *Heterodera schachtii* [71].

## 4. Viroid-induced plant hormone alterations

Viroids are agents that can cause disease by infecting many agriculturally important crop groups (such as potatoes, tomatoes, hops, coconuts, and citrus fruits) and ornamental plants [72–74]. Although viroids cause symptoms similar to viral infections in plants, they have exceptional structural, functional, and evolutionary features. Viroids are RNA with a minute genome varying between 250 and 401 bases, single-stranded, circular, non-protein-coding, and autonomously replicating RNA [75, 76]. They need a host cell to reproduce. Viroids are the smallest highly structured infectious agents that do not have detectable mRNA activity and can reproduce spontaneously in susceptible plant species [77]. Unlike viruses, viroids that do not have protein-coding ability need RNA-dependent RNA polymerases in the host plant to perform the initiation and elongation of viral RNA strands in replication and use cellular pathways for the transport of proliferating viroids [78, 79]. To date, 33 viroid species belonging to eight genera and two families (Avsunviroidae and Pospiviroidae), accepted by the International Committee on Virus Taxonomy (ICTV), have been reported and recorded. Species belonging to these families have important changes in terms of replication mechanisms and where they reproduce in the cell [80, 81]. In viroid infection, symptoms may present as asymptomatic, mildly symptomatic, or severe symptomatic, according to the plant species and the variant of the agent and/or the early infection of the plant. The symptom severity observed in plants varies depending on abiotic factors, such as high temperature and light intensity, that affect viroid replication [72, 82, 83].

Potato spindle tuber viroid (PSTVd) is the first agent defined as viroid in potato (*Solanum tuberosum* L.) viroids [84]. It has been reported to cause intense symptoms in this plant, where it is the main host, and cause serious damage in large production areas [85]. PSTVd is asymptomatic in most plants but has a broad host range (*Solanaceae*, *Asteraceae*, Gesneriaceae, and Lauraceae). Among the most important symptomatic hosts are tomato (*Solanum lycopersicum* L.) [85] and pepper (*Capsicum annuum* L.) [86] plants. Until today, it is known that the resistance to viroids in cultivated plants is limited, and there are no naturally resistant plants. Viroid infection usually results in changes in the cellular functions of the host plant, including defense mechanism, signal transduction, hormone balance, and energy production [77]. Viroids'virulence mechanism is mediated either straightforwardly by their own genome or by single- or double-stranded RNAs (ss/dsRNAs) composed of its genome, and the symptoms that arise systemic infection are caused by small RNAs (sRNA) produced from the viroid, which is unknown to date. It can be a result of accumulation with plant features, either in the organelle where the viroid proliferates or in the cytoplasm thru its effort within the cell [72]. Application of microarray analysis-based studies in viroid-pathogenicity study due to the low proliferation rate of viroids in the model plant A. thaliana [87–89]. It was determined that PSTVd infection caused an important modification in gene expression in one of four different genes in Rutgers tomato cultivar, similar to the result observed in CaLCuV infection in A. thaliana [90].

To date, things have been done about gene-specific events different from host diseases [91]. It is a mediator of plant hormones to signals in plant defense against plant organisms and in response to biological stress [92, 93]. Plant hormones as a whole took part in the regulation of defense; Hormones involved in plant defense, such as salicylic acid (SA), ethylene (ET), and jasmonic acid (JA), are significant in systemic gain in the fight against pathogens [94]. Brassinosteroid (BR-induced systemic defense against biotrophs, appearance as a system from the entity that differs in appearance from the SA mediated SAR and JA/ET from the entity that is different from the mediated disease [95, 96]. Plant hormones such as abscisic acid, auxins, cytokines, plant systems, and plant hormones such as gibberellins are those that have emerged as end-model choices because they alter plant system functioning and morphological responses [97]. Having simple structures and low systems, plant hormones, cell type, plant preference, and as well as organogenesis and apoptosis preference, is in the eye and Estel (Santner). They also play a role in defense against viruses [98]. Viroids, those with non-RNA-coding genomes, also stimulate defensive responses in plants for symptomatic. In its design as a model system, the expression of PSTVd factor and its hosts encoding products involved in viroid defense/stress response, growth propensity, preferred defense/stress response, development, and other functions has been overlooked [98].

In recent years, with the increase in microarray analyzes and transcriptome studies, data on the effect of different signaling pathways in viroid infection have been obtained [99]. SA is identified for its significant function in the plant defense reaction to diverse biotic and abiotic stresses [100, 101]. In many plant species, fungal, bacterial, or viral infection causes local and, to a reduced amount, systemic amassing of endogenous SA [102]. Viral contaminations may activate boosted SA quantities both in infected subordinate leaves that acquire a hypersensitive response (HR) and in uninfected higher leaves that improve systemic acquired resistance (SAR) [103, 104]. Many plant species such as tomatoes, soybeans, and potatoes can have basal SA levels significantly greater than the presence in tobacco and A. thaliana [105]. Application of SA to sensitive tobacco and other plant species infected with tobacco mosaic virus (TMV), Cucumber mosaic virus (CMV), and Potato X virus (PVX) increased resistance to these agents [106]. In addition, SA application in tobacco and potato-PVY interactions decreased viral increase and postponed the onset of disease symptoms [107].

To determine the plant-viroid interactions, it was revealed that there was rising in the expression of the NPR1 gene in the susceptible Rutgers tomato variety 21–30 days after the infection of the tomato with PSTVd (potato spindle tuber viroid). It was confirmed that the expression of both NPR1 and PR1 genes was increased in the PSTVd-infested plant [108]. In the transcriptome investigation of tomato plants infested with PSTVd, it was determined that there were noteworthy modifications in the expression of 93 genes linked to SA biogenesis and 146 genes associated with SA signaling [109]. In this way, it has provided a unique insight into the mechanisms underlying transcript profiling, gene resistance and basal defense mechanism for gene, biotrophy versus necrotrophy, and pathogenicity of vascular and non-vascular pathogens. In this way, genomic technologies have facilitated the systems-wide approach to incorporate distinctive structures in the interfaces of hosts and pathogens [110]. It has been determined that increased SA accumulation and external SA application in *Gynura aurantiaca* plants infested with Citrus exocortis viroid (CEVd)

[111] have a beneficial effect on alleviating the symptoms caused by CEVd [112]. In addition, PR1, PR2, and PR3 gene expression levels were increased in the leaves of the hop plant infested by Citrus bark cracking viroid (CBCVd) and/or Hop latent viroid (HLVd) [113]. In HSVd infected hop plants [114], It is suggested that the regulation of genes encoding SA-bound PR1 and PR4 proteins was not triggered. No important alterations in endogenous SA content are detected in the PSTVd-infested potato cultivars *Solanum laxum* and Désirée [115, 116]. The poor reaction of potato to SA in PSTVd contamination suggests that several still unrevealed salicylates play in signal transduction in potatoes. Gentisic acid (GA), a metabolic derived of SA, observes to accumulate in large amounts, even higher than SA, in tomato and *G. aurantiaca* plants infested with CEVd [117]. It has also been suggested that GA treatments can delay the onset of symptoms and that GA plays as a pathogen-derived signal in addition to SA for the stimulation of plant defense genes in tomatoes [118].

JA plays a significant role in plant growing and growth events such as reproduction, flower and fruit growth, root development, senescence [119], and jasmonates in injury [120], insect damage, and defense against pathogens were determined to be key signals in stress responses [121]. Although JA does not inhibit the systemic increase of viruses, it has been observed that external applications of JA can reduce viral replication [122]. Generally, it has been observed that JA and ET signaling pathways do synergistically in the plant defense mechanism [123], and both JA and SA are necessary for pathogen-associated molecular model (PAMP)-stimulated resistance reactions to potato [123]. While JA supports plant defense in the initial stages of viral contamination, it reduces plant resistance if applied in later stages [124]. In *Nicotiana benthamiana*, both SA and JA were required for systemic resistance to TMV, and inhibiting JA development significantly reduced subsequent SA accumulation, suggesting that the reverse had no effect, proposing that JA was required to motivate SA [125]. Evidence that JA-mediated signaling is involved in plant-viroid interactions in tomato plants infested with PSTVd is provided by genome-wide analyzes [108].

A transcriptome study in hops showed that the synthesis of JA biosynthesis and JA signaling pathways was triggered in HSVd-infected plant leaves [114]. While the expression of the LOX gene was decreased in complete tissues of the hop plant (leaf, flower, and cone) infected with asymptomatic HLVd and severe symptomatic CBCVd in the hops plant, the JAR1 gene revealed small fluctuations in gene expression competed to the plants without viroids, regardless of disease status. Decreases in the expression of the LOX gene in leaf samples of together plant groups indicate that this gene does not contribute significantly to the dwarfing of hops plants infected with CBCVd [113].

In a study on gene expression in the tomato-PSTVd interaction, JAZ and MYC2 genes were identified for JA signaling [109], while MYC2 is the main regulator of the JA signing pathway, which triggers the transcription of initial JA-receptive genes, whereas, in jasmonate-induced transcription, it is found to have a dual role. Although there are different variants of PSTVd used in studies, systemic infection of Rutgers tomato cultivar with PSTVd was confirmed by increased accumulation of JA in plant leaves [126]. Significant amassing of endogenic JA and its precursor cis-OPDA was detected in plants of PSTVd-infected Désirée potato cultivar and *S. laxum* plants [115, 116]. However, no notable modifications in JA and cis-OPDA contented or appearance of related genes were observed in the tubers of infected potato plants, recommending the presence of organ-specific alterations in plant hormone reactions upon potato-PSTVd contamination. Deficiency of JA biosynthesis and signaling can cause male sterility [127] and undersized development of aerial plant parts, particularly in situations where endogenic gibberellins are reduced [128].

Brassinosteroids (BRs) are steroid phytohormones that have been widely studied for their early isolation and characterization [129]. Plant growth and development [130].

They also act in plant resistance by stimulating plant defense against viruses [95]. BRs have active roles in defensive plants in contradiction of several biotic and abiotic stresses [131, 132]. It has been determined that the application of brassinolide (BL) against TMV in tobacco reduces the lesions amount on tobacco leaves and increases resistance to TMV [95]. BL application also reduced symptom development in CMV-infected Arabidopsis [133]. The interaction of BR with other plant hormones is partly related to the ability of BR to confer stress tolerance through synergistic or antagonistic interfaces with additional stress plant hormones such as auxins, JA, gibberellins, SA, and ABA, [134].

When a different PSTVd variant was used on the same tomato cultivars, it caused a rise in the expression of some genes like brassinosteroid-6-oxidase 1, while decreased expression of the CPD gene was observed [108]. It was determined that PSTVd infection did not have a noteworthy influence on endogenic CS in infested tomato plants [126]. The function of BR-mediated signing in plant reactions to viroid interfaces is highly dependent on the host-viroid grouping, and though BRs have a significant part in the regulation of plant immunity, it is not possible to predict the outcome due to the different mechanisms of action of BRs in plant-viroid infection.

The significance of plant hormones in the instruction of plant immunity is clear [135]. There is insufficient information to form a general model of the interaction of plant hormones during plant-viroid infections. Therefore, new research can help improve the product by adding phytohormones aimed at improving plant resistance/tolerance to viroids in the production program at the aquaculture stage.

## 5. The interaction of phytoplasma with plant hormones

Many yellow-type diseases were supposed to be caused by viruses until a group of Japanese researchers in 1967 saw microorganisms similar to animal mycoplasma through electron microscopy in the phloem of diseased plants. Later, these types of bacteria without a cell wall were called mycoplasma-like organisms (MLOs) [136]. In subsequent years, MLOs were named "*Candidatus phytoplasma*" with DNA-specific amplification and sequencing methods [137]. Phytoplasma 16S rRNA genes with sufficient diversity are greatly preserved and are used in classifications of phytoplasmas [137–140].

Despite having one of the smallest genomes among living organisms, phytoplasmas encode compound metabolic functions that enable them to network with their plant hosts [141, 142]. Phytoplasmas are of different sizes and shapes, they live and reproduce in environments with equal osmotic pressure provided by plant phloem and insect fluids, and it is still not possible to reproduce in *in vitro* conditions. Phytoplasmas are plant diseases that spread violently and rapidly and can increase metabolic activities, such as plant shoot production, flower shape, and color change in their hosts. They also cause severe decline and death in infected plants [143, 144]. Plants infected with phytoplasmas often show symptoms indicating that their growth regulators are severely affected. The characteristic symptoms are vigor and phyllodes of flowers, sterility, loss of apical dominance leading to witches' broom formation and proliferation of axillary buds, abnormal internode elongation, and general stunting [145, 146].

Phytoplasmas are transmitted by families of *Cicadellidae*, *Cixidae*, *Psyllidae*, *Delphacidae,* and *Derbidae* [147]. It has been determined that phytoplasmas are transmitted with transovarial and seeds in some insect and plant hosts [148–157]. Phytoplasmas are also efficiently spread by vegetative propagation like micropropagation parts, grafting, and cuttings [158–160].

Phytoplasma diseases were previously thought to affect the growth regulators of plant hosts. It is now recognized that the pathogenicity of phytoplasmas includes certain effector proteins that have diverse effects on growth and other aspects of plant life. Phytoplasma effector proteins activate physiological events in cells that control symptom development in the plant. Such physiological changes in plants result from the effect of phytoplasma on hormonal, nutritional, developmental processes, and stress and the interaction of these changes with each other [161].

Plant hormones are small molecules that have important regulatory roles in plant growth, survival, and immunity [162–172]. Plant hormones also trigger plant defense mechanisms against pathogen attacks [173, 174]. Salicylic acid (SA), jasmonic acid (JA), and ethylene are known to be involved in plant immune reactions. Although there are not many studies on other plant hormones such as cytokinins, abscisic acid, auxins, brassinosteroids, gibberellic acid, and peptide hormones, it is known that these hormones also take place in plant defense systems. These hormones are important for plant defense mechanisms in pathogen attack, and the cooperation between hormones is preserved in plant species. Activation and regulation of hormones against pathogen attack depend on plant host-phytoplasma interaction [173–176].

SA is the plant hormone used against pathogens in plant defense and enables the Pathogenesis Related Protein (PR) protein genes to be activated. The PR-1, PR-2, and PR-5 proteins are used as molecular markers for monitoring SA-dependent gene expression [176–179]. Genes expressed in the presence of SA are up-regulated in phytoplasma-infected plants. In a study, PR proteins were up-regulated in phytoplasma-infected seedlings of garland chrysanthemum and Mexican lime infected with "Ca. P. aurantifolia." When the mulberry phloem sap was infected with phytoplasma, the expression of PR-1 protein was increased [180–182]. SA-hormones activities are increased in the grapevine infected with the phytoplasma "Ca. P. solani" and flavescence dorée phytoplasma [183–187].

JA is an important plant hormone involved in the development processes in which stress reactions such as insect attack, injury, and drought are triggered [188, 189]. Significant variation in gene expression by JA signaling has been reported in the phytoplasma infected tobacco (infected with the "Ca. Perrisia mali"), grapevine (infected with bois noir; up-regulated PR-6 gene), tomato (infected with "Ca. P. solani" strain C; up-regulation of the PR-6 gene PIN2), coconut (infected with yellow decline phytoplasma; up-regulation of the PR-3 and PR- 10) [186, 190].

Ethylene is a minor gas hormone that regulates the expansion of organs such as leaves, flowers, and fruits in plants and is considered to have an effect on activating the plant defense system against pathogens [173, 191]. Ethylene signaling is generally in synergy with JA signaling and provides increased resistance to pathogen attack by the expression of certain defense genes [173]. Differentiation of gene expression by ethylene signaling has been reported in the phytoplasma infected Mexican lime (infected with "Ca. *P. aurantifolia*"), grapevine (infected with "Ca. *P. solani*" and "Ca. *P. asteris*"), *Paulownia fortunei* (paulownia witches' broom phytoplasma) [183, 190, 192–195].

There are some researches that have revealed the plant-pathogen interactions of Indole-3-acetic acid (IAA) or auxin signaling in disease development [171, 196]. In one study, it was revealed that there was a significant differentiation in the level of IAA in phytoplasma-infected Mexican lime [197].

Cytokinins are an important group of plant hormones that involve in plant growth and development [161]. In a study, it is thought that the elevation of cytokinin in the flowers of *Catharanthus roseus* plants is caused by phytoplasma infection [198].

Genetic studies in recent years have been promising in understanding the disease capabilities of phytoplasmas, the plant-phytoplasma interaction, and the control of phytoplasma diseases in plants. These studies have also revealed the hormonal responses of plants to phytoplasma diseases and the resistance mechanisms they have developed in the fight against the disease. However, it is obvious that there are still many unknowns waiting to be clarified in terms of the biological structure of phytoplasmas and the hormonal interactions of phytoplasma infected plants. When phytoplasma diseases and plant resistance mechanisms are understood, agricultural production with disease control will be possible over the next half-century.

## 6. Phytohormone-producing plant growth-promoting rhizobacteria

Plants have a greatly complicated defense system akin to the animal immune system. Unknown molecules or signals are recognized by plants from their own injured cells and their immune responses by triggering against the invader [199, 200]. Plant hormones play a part in the regulation of plant growth. They have important functions in the setting of immune reactions against pathogens, insects, and beneficial microorganisms. Signaling pathways are interrelated with a complex network. Plants have a huge organizing potential to adapt to the biotic environment rapidly. They employ inadequate resources for growing and trigger their immune system against attack by insects or pathogens.

Plant pests possess several mechanisms to operate the plant's hormone signaling interaction to overcome host immunity. Beneficial root-colonizing microorganisms can organize the hormone-regulated immune signaling system to keep a continued mutualistic life cycle [201].

Efficacious pathogens and insects may vigorously attack the plant immune system to launch a continued relationship. On the other hand, beneficial associations between plants and microbes are rarely found in nature. Beneficial microorganisms that are found in the roots or the rhizosphere considerably progress in plant growth. Because beneficial microorganisms are primarily recognized by plants as possible attackers, thus plant immune system is become activated; they also support the plant to withstand biotic or abiotic stress [202].

Beneficial plant-microbe associations are highly organized with coordination of plant immune responses that are significant for plant development and survival. The best-studied mutualistic microorganisms are mycorrhiza fungi, plant-growth-promoting rhizobacteria, fungi, and Rhizobium bacteria [203, 204].

Rhizobium bacteria fix atmospheric nitrogen for the plant by inducing the development of symbiotic constructions (nodules) in the roots of legumes [201].

Beneficial PGPRs are in close relations with plant roots. They may belong to several bacterial genera such as *Pseudomonas*, *Bacillus*, *Azotobacter*, *Burkholderia*, *Azospirilium*, *Paenibacillus.* Rhizobacteria may enhance plant growth by generating phytostimulators or by overwhelming pathogen microorganisms [205, 206]. They also produce phytohormones. The six major classes of phytohormone produced by PGPR may induce plant growth, initiate plant immunity, sustain stress tolerance, and support plant maturity for fruiting and seedling [207]. From this point of view, phytohormone-producing PGPR covers the role and significance of sustainable agriculture. Some of them may regulate plant physiology by mimicking the synthesis of plant hormones. Plant-growth-promoting rhizobacteria (PGPRs) cause a series of chemical changes in soil by increasing minerals and nitrogen. They are applied to crops for growth enhancement and disease control [208, 209]. Various stress types of salt, including chilling, nutritional stress, pathogen infection, heat, metals, and wounding, may cause damage to crops [210]. The use of PGPR may affect plant hormone signaling pathways by generating gibberellin, auxins, ABA, and cytokinins or by intervening plant ethylene quantities by forming ACC deaminase reduced the damage to plants that follow in stress situations by definite mechanisms at sites where stress is a chief restriction [211]. Salicylic acid is a significant plant hormone vital for the immune reaction to pathogens. SA has an essential role in disease resistance signaling [212]. The SA reaction pathway is typically effective against pathogens [213]. SA is a phenolic compound that may be synthesized from the primary metabolite by systemically induced disease resistance responses [214]. Plants are involved in complex signal transduction pathways [215].

Jasmonic acid (JA) is a plant hormone vital for the immune reaction to pathogens. JA was initially isolated from plants of the *Jasminum* spp. and other fruits. It can modulate the ripening of fruits, root growth, pollen production, tendril coiling, and plant resistance to pathogens. JA concentrations in uninduced plant tissues are mostly very minor in plant species [216, 217]. JA biosynthesis is controlled by various elicitors such as wounding [210, 216, 218]. The expressions of numerous genes, with lipoxygenase (LOX) and AOS, were boosted by exogenous application of JA [219–222] and connected with an amplified level of endogenous JA [223, 224].

Ethylene ($C_2H_4$) is a simple gaseous hydrocarbon that has special effects on plant development and growth [225]. A chief component of resistance signals is constructed thru numerous plant-pathogen interfaces. It is a significant modulator of plant immunity [226–228]. Ethylene has an important potential in the growth of the plant. Although greatest commonly associated with ripening, ethylene is a regulator of seed germination, leaf and petal abscission, seedling growing, stress, organ senescence, and pathogen reactions.

Abscisic acid is a 15-C weak acid that was identified in the early 1960s. It is a development inhibitor amassing in abscising cotton fruit ("abscisin II") and sycamore trees stimulated to develop dormant ("dormin") [229–232]. ABA has been adjusting some plant development and growth features such as seed dormancy, embryo maturation, elongation, germination, cell division and floral initiation, and reactions to ecological stresses like cold, drought, salinity, pathogen incident, and UV radiation. The presence of ABA in abscising organs reveals its potential in stimulating senescence and/or stress reaction. Young tissues have great ABA levels, and ABA-deficient mutant plants are highly undersized in part since their capability in reducing transpiration [233]. Recent studies have greatly concentrated on understanding of ABA signaling mechanism [234]. ABA is a significant modulator of the plant immune signaling complex. Besides this, it has an important role in growth and modification to abiotic stress such as salinity and drought stress [235–237]. Plant immunity is suppressed by ABA signaling using the SA-dependent defense mechanism [238–241].

Auxins play an essential role in plant development [242]. Microorganisms may produce auxins or influence auxin signaling in the host [243, 244]. Auxin signaling may suppress SA quantities and signaling [245]. Many pathogens changed ways to achieve auxin-mediated destruction of SA to improve the sensitivity of the host [246].

The GAs encompasses many diterpenoid carboxylic acids, common in higher plants as endogenous development regulators that involve in promoting organ growth and growing variations. These complexes are also constructed by several bacteria, plants, and fungi species. The role of GAs in microorganisms has only lately been researched. Biologically functioning GAs concentration is firmly controlled and is moderated by various developing and environmental signals. The latest investigation has focused on the expression of the genes that encode the dioxygenases plays a role in biosynthesis and inactivation. GAs influence plant development by regulating the deprivation of development inhibitory DELLA proteins [247]. They tightly play a role in plant immune signaling [248]. MAMPs are produced by soil-borne microorganisms [249, 250]. Because of the abundance of soil rhizosphere inhabiting microorganisms, immune signaling in plant roots should be controlled frequently [251]. Hormone-regulated responses may affect the composition of the native microflora of the rhizosphere [252, 253]. Beneficial soil microorganisms use strategies that rely on hormone-adjusted immune reactions that are activated in the roots on preliminary detection [254].

## 7. Virus-induced-phytohormone remodeling in plant tissues

Plant hormones ethylene (ET), salicylate (SA), jasmonate (JA), and abscisic acid (ABA) primarily regulate synergistically or antagonistically against pathogens [251]. Plants respond to pathogen invasion by the hypersensitive response (HR), R-mediated, and systemic acquired resistance (SAR). The signaling of systemic acquired resistance (SAR) depends on the discernment of ethylene and salicylic acid [255] synergistically [256] or independently and antagonistically [257]. Phytohormones and precursors, such as salicylic acid (SA) and aminocyclopropane-1-carboxylic acid (ACC), increased antioxidant enzyme activities and defense-associated genes related to SAR [257].

Hypersensitive response of plant cell death reinforces systemic acquired resistance to prevent pathogen movement. The genes are generally associated with plants expressing systemic acquired resistance and hypersensitive response induce ROS (Reactive Oxygen Species) MAP kinase activation, protein phosphorylation, changes in the ion fluxes, and phytohormones as the first responses to stress conditions. The intersection of emerging signals causes reprogramming of defensive metabolites [258, 259]. Integration of abscisic acid (ABA) and salicylic acid (SA) signals resulting from pathogen-related (PR) gene expression during pathogen invasion induce defense mechanism and increase expression of genes associated with resistance [103]. Otherwise, different SA signaling pathway is responsible for PR gene expression independently and related alternative oxidase (AOX) induction [260]. Expression of Cucumber mosaic virus protein cmv2b inhibits SA-mediated resistance by inducibility of AOX (alternative oxidase) [261]. Relationship between AOX and virus resistance, explained by disruption of redox signaling in the mitochondrion [262]. SA modulates the opening of mitochondrial permeability transition pore, and alternative oxidase (AOX) limits the systemic spread of the viruses [263].

Expressed transcriptional factors regulate responsive defense hormone in plants during pathogen invasion. The NAC transcription hormone signals [264] response to pathogen infection as well as to stress conditions. TMV-P0 inoculated, and plant defense-related hormone-treated Capsicum plants overexpress the CaWRKYd gene. Silencing of CaWRKYd affects HR lesion formation [265]. WRKY transcription factors identified as abscisic acid (ABA) signaling regulators. WRKY40 regulates the expression of genes responsible for ABA production and modifies the transcription of numerous defense-related genes.

Callose accumulation and stomatal closure limiting virus movement are the first responses to stress conditions regulated by the plant stress hormone ABA [266]. Exogenous application of ABA increases resistance to tobacco mosaic virus [267] and tobacco necrosis virus infection [268]. The abscisic acid (ABA) acts both resistance and susceptibility side against virus diseases related to different diseases such as Cucumber mosaic virus (CMV) and Bamboo mosaic virus (BaMV) [251]. ABA treatment increases antiviral defense responses of plants and contributes antiviral disease resistance [269]. ABA postponements the multiplication of TMV in invaded leaves and prevents systemic movement. WRKY8 prevents virus infection regulated by ABA signaling enhances immune responses against viruses. Another transcription factor ABA-responsive MYB gene identified from the avian myeloblastosis virus was widely distributed in plants [270]. MYBs interrelate with phytohormones in plant defense against virus disease and serve as a molecular connection to integrate abscisic acid (ABA) and salicylic acid (SA) signals and enhance disease resistance induction of PR genes during *Pseudomonas syringae* invasion [271]. MYB1 performs in the crosstalk between phytohormones and defense responses to regulate the SA-dependent defense responses against TMV [257].

Ethylene (ET) plays a complex role in resistance against pathogens, and the scheduling of ethylene application affects plant defense against viral infection [251, 257]. ET contributes to virus susceptibility and plays a negative role in anti-TMV-cg defense. Ethylene-mediated MYB4L signaling pathway affects the susceptibility to TMV as a consequence of the SA-independent resistance mechanism [257]. Aminocyclopropane-1-carboxylic acid (ACC) treatment increased the accumulation of TMV-cg systemically [267] and Turnip mosaic virus (TuMV) [272]. On the contrary, ACC, the precursor of ethylene, accumulates around necrotic lesions of TMV-infected plants, and SAR depends on the perception of ethylene [273]. Supplementation of the ACC had a suppressive influence on the replication of the white clover mosaic virus [122]. ACC-pretreatment increases the transcription of MYB4L and stimulates ethylene production during viral infection to enhance TMV resistance [257]. Overexpression of ET-response transcription factor (ERF) enhances resistance to tobacco mosaic virus infection preventing systemic spread of the virus independently exogenous application of salicylic acid, jasmonic acid, or ethylene [274].

Another evidence of phytohormone-mediated resistance to viruses is viral silencing suppressors (VSRs) [261]. VSRs regulate to integrate connections between hormone signaling and subcellular organization [275]. Most viruses encode silencing suppressors (VRSs) to suppress SA-related defense responses [276]. SA-dependent signaling pathway leads the virus resistance [251]. RNA reliant on RNA polymerase (RdRp) open reading frame (ORF) is responsible for RNA silencing suppressors to inhibit SA signaling and ABA [277]. CMV 2b expression inhibits the salicylic acid (SA)- mediated defense response. Virus-encoded suppressor of RNA silencing protein can constrain to overcome SA-mediated resistance mechanism [278]. An exogenous SA application capable of induction RNA silencing antiviral defense response to VSR protein [261]. Systemic movement of ToRSV was reduced with SA pretreatment. SA signaling plays a role restriction of systemic infection but not RNA silencing [279]. Conversely, Alamillo et al. [280] reported that the SA signaling system enriches the function of the RNA silencing system in tobacco. SA-inducible RNA-dependent RNA polymerases (RDRs)-mediated RNA silencing mechanism and the AOX-regulated pathway enhanced resistance to tobamoviruses independent by PR genes [263, 281]. Exogenous SA application suppressed replication of Potato virus Y in tobacco [282], and RNA silencing mechanism associates plant immunity against viral pathogens [283].

Virus spread is commonly limited by the siRNA microRNA system at the transcriptional or post-transcriptional level [251]. The SA-dependent overlaps between hormones and siRNA, miRNA [262, 284], RNA silencing system [256, 262, 284] (Yu et al. 2003), JA- mediated (VSR) proteins [276], CMV-2b-VSR interferes [261] were evidenced in various studies.

ARFs are the key factors in auxin signaling factors that interact with the helicase domain of TMV replicase [285]. TMV replicase-Aux/IAA interactions increase virus pathogenicity in tissues of tomato plants where Aux/IAA proteins accumulate [286]. Repression of the auxin receptor increased rice susceptibility to Rice black-streaked dwarf virus (RBSDV), and Rice stripe virus (RSV) expression of auxin signaling genes changed in response to virus infection [287, 288]. Tobacco mosaic virus (TMV) disrupts auxin/indole acetic acid (Aux/IAA) transcriptional regulators and access to the phloem for systemically spreading [289]. Tomato spotted wilt virus (TSWV) manipulates auxin signaling by interruption of Auxin Response Factor

(ARFs) to optimize their own replication and spread [290, 291]. Viral suppressor HCPro component of Tobacco vein banding mosaic virus (TVBMV) triggers auxin biosynthesis and an increase in auxin accumulation in plants [281].

In the last century, endogenous cytokinins and pathogen interactions in plants propounded [292]. Cytokinin levels confer resistance against White clover mosaic virus [122], Tobacco necrosis virus (TNV) [293], Potato virus Y [294]. Cytokinin may modulate the redox signaling system and trigger defense reactions [278]. Cytokinin accumulation suppresses virus-induced necrotic symptoms and increases ascorbate peroxidase and catalase activities, ascorbate levels, and H2O2-scavenging capacity [293]. White Clover Mosaic Potexvirus infection decreases the level of free bases but increases the accumulation of nucleotides. Despite the cytokinin concentration being the same for both control and infected leaves, the distribution of the specific cytokinin groups in the virus-invaded leaves had changed significantly [122].

Researchers interested in the relationship between virus accumulation and gibberellic acid (GA) and the earlier findings demonstrated that virus content was not reduced by GA [295]. However, virus infection affects GA biosynthesis and accumulation. Rice dwarf virus (RDV) infection reduces GA biosynthesis in rice [291], tomato yellow leaf curl Sardinia virus (TYLCSV) in S. lycopersicum [296]. Tobacco mosaic virus in tobacco [297] and CMV significantly reduces the content of jasmonic acid (JA), indole acetic acid (IAA), salicylic acid (SA), and gibberellic acid (GA3) in cucumber plants [298]. Pathogen invasion induces BR biosynthesis and signaling pathways in both resistant and susceptible plants. The resistance against the Rice stripe virus (RSV) can be enhanced by Brassinosteroids (BR) and jasmonic acid (JA) signaling, and RSV invasion is suppressed by the endogenous BR levels [299].

## Conflict of interest

The authors declare no conflict of interest.

## Author details

Refik Bozbuga[1*], Bekir Bulent Arpaci[2†], Selman Uluisik[3†], Pakize Gok Guler[4†], Hatice Nilufer Yildiz[4†] and Songul Yalcin Ates[5†]

1 Faculty of Agriculture, Plant Protection Department, Eskisehir Osmangazi University, Odunpazarı, Eskisehir, Turkey

2 Faculty of Agriculture, Department of Horticulture, Saricam, Cukurova University, Adana, Turkey

3 Burdur Food Agriculture and Livestock Vocational School, Burdur Mehmet Akif Ersoy University, Burdur, Turkey

4 Biological Control Research Institute, Adana, Turkey

5 İzmir Directorate of Agricultural Quarantine, İzmir, Turkey

*Address all correspondence to: refik.bozbuga@ogu.edu.tr

† Authors contributed equally to this work. Hormone-plant interactions induced by parasitic nematodes (by R Bozbuga), plant hormones (by S Uluisik), virus (by BB Arpaci), viroid (by PG Guler), bacteria (by HN Yildiz), and phytoplasma (by SY Ates) are written in this book chapter.

## References

[1] Aires A, Carvalho R, Matos M, Carnide V, Silva AP, Gonçalves B. Variation of chemical constituents, antioxidant activity, and endogenous plant hormones throughout different ripening stages of highbush blueberry (*Vaccinium corymbosum* L.) cultivars produced in Centre of Portuga. Journal of Food Biochemistry. 2017;**41**(6): e12414

[2] Li J, Liu B, Li X, Li D, Han J, Zhang C, et al. Exogenous abscisic acid mediates berry quality improvement by altered endogenous plant hormones level in "Ruiduhongyu" grapevine. Frontiers in Plant Science. 2021;**12**:739964

[3] Rohrmann J, Tohge T, Alba R, Osorio S, Caldana C, McQuinn R, et al. Combined transcription factor profiling, microarray analysis and metabolite profiling reveals the transcriptional control of metabolic shifts occurring during tomato fruit development. Plant Journal. 2011;**68**(6): 999-1013. DOI: 10.1111/j.1365-313X.2011.04750.x

[4] Xu Y, Li K, Zhu K, Tian Y, Yu Q, Zhang WZ. Effect of exogenous plant hormones on agronomic and physiological performance of a leaf early-senescent rice mutant osled. Plant Grownth Regulation. 2020;**92**:517-533

[5] Petersson SV, Johansson AI, Kowalczyk M, Makoveychuk A, Wang JY, Moritz T, et al. An auxin gradient and maximum in the Arabidopsis root apex shown by high-resolution cell-specific analysis of IAA distribution and synthesis. Plant Cell. 2009;**21**:1659-1668. DOI: 10.1105/tpc.109.066480

[6] Li S-W, Xue L, Xu S, Feng H, An L. Mediators, genes and signaling in adventitious rooting. Botanical Review. 2009;**75**(2):230-247

[7] Ljung K, Hull AK, Celenza J, Yamada M, Estelle M, Normanly J, et al. Sites and regulation of auxin biosynthesis in Arabidopsis roots. Plant Cell. 2005;**17**(1090):1104

[8] Guan L, Tayengwa R, Cheng ZM, Peer WA, Murphy AS. Auxin regulates adventitious root formation in tomato cuttings. BMC Plant Biology. 2019;**19**: 435. DOI: 10.1186/s12870-019-2002-9

[9] Bouzroud S, Gouiaa S, Hu N, Bernadac A, Mila I, Bendaou N, et al. Auxin response factors (ARFs) are potential mediators of auxin action in tomato response to biotic and abiotic stress (Solanum lycopersicum). PLoS One. 2018;**13**(2):e0193517. DOI: 10.1371/journal.pone.019351

[10] Guo Y, Jiang Q, Hu Z, Sun X, Fan S, Zhang H. Function of the auxin-responsive gene TaSAUR75 under salt and drought stress. The Crop Journal. 2018;**6**:181-190. DOI: 10.1016/j.cj.2017.08.005

[11] De Jong M, Wolters-Arts M, García-Martínez JL, Mariani C, Vriezen WH. The Solanum lycopersicum AUXIN RESPONSE FACTOR 7 (SlARF7) mediates cross-talk between auxin and gibberellin signalling during tomato fruit set and development. Journal of Experimental Botany. 2011;**62**:617-626

[12] Goetz M, Hooper LC, Johnson SD, Rodrigues JC, Vivian-Smith A, Koltunow AM. Expression of aberrant forms of AUXIN RESPONSE FACTOR8 stimulates parthenocarpy in Arabidopsis and tomato. Plant Physiology. 2007;**145**(2):351-366. DOI: 10.1104/pp.107.104174

[13] Shin J-H, Mila I, Liu M, Rodrigues MA, Vernoux T, Pirrello J, et al. The RIN-regulated small auxin-up RNA SAUR69 is involved in the

unripe-to-ripe phase transition of tomato fruit via enhancement of the sensitivity to ethylene. The New Phytologist. 2019;**222**:820-836

[14] Chen K, Li GJ, Bressan RA, Song CP, Zhu JK, Zhao Y. Abscisic acid dynamics, signaling, and functions in plants. Journal of Integrative Plant Biology. 2019;**62**:25-54

[15] Kou X, Yang S, Chai L, Wu C, Zhou J, Liu Y, et al. Abscisic acid and fruit ripening: Multifaceted analysis of the effect of abscisic acid on fleshy fruit ripening. Scientia Horticulturae. 202;**281**:109999. DOI: 10.1016/j.scienta.2021.109999

[16] Xie EJ, Pang X, Chen H, Cai W. Exogenous application of abscisic acid to shoots promotes primary root cell division and elongation. Plant Science. 2020;**292**:110385. DOI: 10.1016/j.plantsci.2019.110385

[17] Vishwakarma K, Upadhyay N, Kumar N, Yadav G, Singh J, Mishra RK, et al. Abscisic acid signaling and abiotic stress tolerance in plants: A review on current knowledge and future prospects. Frontiers in Plant Science. 2017;**8**:161

[18] Dekkers BJW, Schuurmans JAMJ, Smeekens SCM. Interaction between sugar and abscisic acid signalling during early seedling development in Arabidopsis. Plant Molecular Biology. 2008;**67**:151-167

[19] Mathan J, Singh A, Ranjan A. Sucrose transport in response to drought and salt stress involves ABA-mediated induction of OsSWEET13 and OsSWEET15 in rice. Physiologia Plantarum. 2020;**171**:620-637. DOI: 10.1111/ppl.13210

[20] Gietler M, Fidler J, Labudda M, Nykiel M. Abscisic acid—Enemy or savior in the response of cereals to abiotic and biotic stresses? International Journal of Molecular Sciences. 2020;**21**:4607. DOI: 10.3390/ijms21134607

[21] Cao FY, Yoshioka K, Desveaux D. Journal of Plant Research. 2011;**124**:489

[22] Mine A, Berens ML, Nobori T, Anver S, Fukumoto K, Winkelmüller TM, et al. Pathogen exploitation of an abscisic acid- and jasmonate-inducible MAPK phosphatase and its interception by Arabidopsis immunity. Proceedings of the National Academy of Sciences of the United States of America. 2017;**114**:7456-7461

[23] Bakshi A, Shemansky JM, Chang C, Binder BM. History of research on the plant hormone ethylene. Journal of Plant Growth Regulation. 2015;**34**:809-827. DOI: 10.1007/s00344-015-9522-9

[24] Liu Y, Tang M, Liu M, Su D, Chen J, Gao Y, et al. The molecular regulation of ethylene in fruit ripening. Small Methods. 2020;**14**:1900485. DOI: 10.1002/smtd.201900485

[25] Poór P, Nawaz K, Gupta R, et al. Ethylene involvement in the regulation of heat stress tolerance in plants. Plant Cell Reports. 2021;**23**(3):1031. DOI: 10.1007/s00299-021-02675-8

[26] Freitas VS, de Souza Miranda R, Costa JH, de Oliveira DF, de Oliveira Paula S, de Castro Miguel E, et al. Ethylene triggers salinity tolerance in maize genotypes by modulating polyamine catabolism enzymes associated with H2O2 production. Environmental and Experimental Botany. 2018;**145**:75-86

[27] Xu L, Xiang G, Sun Q, Ni Y, Jin Z, Gao Z, et al. Melatonin enhances salinity tolerance by promoting MYB108A-mediated ethylene biosynthesis in grapevines. Horticulture Research. 2019;**6**:114

[28] Wang Y, Diao P, Kong L, Yu R, Zhang M, Zuo T, et al. Ethylene

enhances seed germination and seedling growth under salinity by reducing oxidative stress and promoting chlorophyll content via ETR2 pathway. Frontiers in Plant Science. 2020;**11**:1066. DOI: 10.3389/fpls.2020.01066

[29] Wang Y, Jiang H, Mao Z, Liu W, Jiang S, Xu H, et al. Ethylene increases the cold tolerance of apple via the MdERF1B–MdCIbHLH1 regulatory module. The Plant Journal. 2021;**106**: 379-393. DOI: 10.1111/tpj.15170

[30] Denancé N, Sánchez-Vallet A, Gofner D, Molina A. Disease resistance or growth: The role of plant hormones in balancing immune responses and fitness costs. Frontiers in Plant Science. 2013;**4**:155

[31] Helliwell EE, Wang Q, Y. Yang ethylene biosynthesis and signaling is required for rice immune response and basal resistance against Magnaporthe oryzae infection. Molecular Plant-Microbe Interactions. 2016;**29**:831-843

[32] Sun X, Yu G, Li J, Liu J, Wang X, Zhu G, et al. AcERF2, an ethylene-responsive factor of Atriplex canescens, positively modulates osmotic and disease resistance in *Arabidopsis thaliana*. Plant Science. 2018;**274**:32-43. DOI: 10.1016/j.plantsci.2018.05.004

[33] Hooley R. Gibberellins: Perception, transduction and responses. Plant Molecular Biology. 1994;**26**:1529-1555

[34] Rai RK, Tripathi N, Gauntam D, Singh P. Exogenous application of ethrel and gibberellic acid stimulates physiological growth of late planted sugarcane with short growth period in sub-tropical India. Journal of Plant Growth Regulation. 2017;**36**:472-486

[35] Salachna P, Mikiciuk M, Zawadzińska A, Piechocki R, Ptak P, Mikiciuk G, et al. Changes in growth and physiological parameters of ×Amarine following an exogenous application of gibberellic acid and methyl jasmonate. Agronomy. 2020;**10**:980. DOI: 10.3390/agronomy10070980

[36] Miceli A, Vetrano F, Moncada A. Effects of foliar application of gibberellic acid on the salt tolerance of tomato and sweet pepper transplants. Horticulturae. 2020;**6**:93. DOI: 10.3390/horticulturae604009

[37] Moumita J, Biswas PK, Nahar K, Fujita M, Hasanuzzaman M. Exogenous application of gibberellic acid mitigates drought-induced damage in spring wheat. Acta Agrobotanica. 2019;**72**(2):1776. DOI: 10.5586/aa.1776

[38] Zhu Z, Ding Y, Zhao J, et al. Effects of postharvest gibberellic acid treatment on chilling tolerance in cold-stored tomato (*Solanum lycopersicum* L.) fruit. Food and Bioprocess Technology. 2016;**9**:1202-1209. DOI: 10.1007/s11947-016-1712-3

[39] Navarro L, Bari R, Achard P, Lisón P, Nemri A, Harberd NP, et al. DELLAs control plant immune responses by modulating the balance of jasmonic acid and salicylic acid signaling. Current Biology. 2008;**18**(9):650-655. DOI: 10.1016/j.cub.2008.03.060

[40] Li X, Liu W, Li B, Liu G, Wei Y, He C, et al. Identification and functional analysis of cassava DELLA proteins in plant disease resistance against cassava bacterial blight. Plant Physiology and Biochemistry. 2018;**124**:70-76

[41] Wang W, Jin N, Mo X, et al. Exogenous gibberellin GA3 enhances defense responses in Rice to the Brown Planthopper Nilaparvata lugens (Stål). Journal of Plant Biology. 2021;**64**:379-387. DOI: 10.1007/s12374-020-09271-5

[42] Sato K, Kadota Y, Shirasu K. Plant immune responses to parasitic nematodes. Frontiers in Plant Science. 2019;**10**:1165. DOI: 10.3389/fpls.2019.01165

[43] Bird DM, Williamson VM, Abad P, McCarter J, Danchin EGJ, Castagnone-Sereno P, et al. The genomes of root-knot nematodes. Annual Review of Phytopathology. 2009;**47**:333-351

[44] Perry RN, Moens M, Starr JL. Root-knot Nematodes Root-knot Nematodes, Oxfordshire. UK: CAB International; 2010

[45] Bozbuga R, Lilley CJ, Knox JP, Urwin PE. Host-specific signatures of the cell wall changes induced by the plant parasitic nematode, *Meloidogyne incognita*. Scientific Reports. 2018;**8**:17302. DOI: 10.1038/s41598-018-35529-7

[46] Chinnapandi B, Bucki P, Fitoussi N, Kolomiets M, Borrego E, Braun, et al. Tomato SlWRKY3 acts as a positive regulator for resistance against the root-knot nematode Meloidogyne javanica by activating lipids and hormone-mediated defense-signaling pathways. Plant Signaling & Behavior. 2019;**14**(6):1601951. DOI: 10.1080/15592324.2019.1601951

[47] Hussey RS, Mims CW. Ultrastructure of feeding tubes formed in gient-cells induced by root-knot nematode Meloidogyne incognita. Protoplasma. 1991;**162**:99-107. DOI: 10.1007/BF02562553

[48] Bozbuga R. Characterisation of cell walls at the feeding site of Meloidogyne incognita [PhD thesis]. Leeds: University of Leeds; 2017

[49] Gheysen G, Mitchum MG. Phytoparasitic nematode control of plant hormone pathways. Plant Physiology. 2019;**179**:1212-1226. DOI: 10.1104/pp.18.01067

[50] Karssen G, Moens M. Root-knot nematodes. In: Perry RN, Moens M, editors. Plant Nematology. Oxfordshire: CABI. 2006. pp. 59-90

[51] Hutangura P, Mathesius U, Jones MGK, Rolfe BG. Auxin induction is a trigger for root gall formation caused by root-knot nematodes in white clover and is associated with the activation of the flavonoid pathway. Australian Journal of Plant Physiology. 1999;**26**:221-231. DOI: 10.1071/PP98157

[52] Glazer I, Orion D, Apelbaum A. Interrelationships between ethylene production, gall formation, and root-knot nematode development in tomato plants infected with Meloidogyne javanica. Journal of Nematology. 1983;**15**:539-544

[53] De Smet I, Lau S, Voss U, Vanneste S, Benjamins R, Rademacher EH, et al. Bimodular auxin response controls organogenesis in Arabidopsis. Proceedings of the National Academy of Sciences of the United States of America. 2010;**107**:2705-2710. DOI: 10.1073/pnas.0915001107

[54] Kyndt T, Goverse A, Haegeman A, Warmerdam S, Wanjau C, Jahani M, et al. Redirection of auxin flow in *Arabidopsis thaliana* roots after infection by root-knot nematodes. Journal of Experimental Botany. 2016;**67**:4559-4570. DOI: 10.1093/jxb/erw230

[55] Bari R, Jones JD. Role of plant hormones in plant defence responses. Plant Molecular Biology. 2009;**69**(4):473-488

[56] Kim SG, Wu J, Wang Y, White EE, Choi YW, Kim KK, et al. Effect of phytohormones and chemical inhibitors on pathogenesis-related genes identified by differential hybridization in rice suspension culture cells. Plant Pathology Journal. 2010;**26**(4):386-393

[57] Smith CJ. Tansley review no. 86 accumulation of phytoalexins: Defence mechanism and stimulus response system. The New Phytologist. 1996;**132**(1):1-45

[58] Dixon RA, Paiva NL. Stress-induced phenylpropanoid metabolism. Plant Cell. 1995;**7**(7):1085-1097

[59] Wen-kun H, Jing-wen Y, Jian-ping J, De-liang P. Effects of plant hormones on the establishment and development of plant parasitic nematodes' feeding sites. Biotechnology Bulletin. 2021;**37**(7): 56-64

[60] Goverse A, Bird D. The role of the plant hormones in nematode feeding cell formation. In: Jones J, Gheysen G, Fenoll C, editors. Genomics and Molecular Genetics of Plant-Nematode Interactions. Dordrecht: Springer Science+Business Media; 2011. pp. 325-347

[61] Kant U, Ramani V. Insect induced plant galls in tissue-culture. Proceedings of the Indian Academy of Sciences—Animal Sciences. 1990;**99**:257-265

[62] Oosterbeek M, Lozano-Torres JL, Bakker J, Goverse A. Sedentary plant-parasitic nematodes alter auxin homeostasis via multiple strategies. Frontiers in Plant Science. 2021;**12**: 668548. DOI: 10.3389/fpls.2021.668548

[63] Bozbuga R, Uluisik S. In: Yildiz M, Ozgen Y, editors. Molecular Host-Nematode Interactions and Tuber Development, Solanum tuberosum—A Promising Crop for Starvation Problem. London, UK: IntechOpen; 2021. DOI: 10.5772/intechopen.98976. Available from: https://www.intechopen.com/chapters/77539

[64] Ribeiro DM, Araujo WL, Fernie AR, Schippers JH, Mueller-Roeber B. Translatome and metabolome effects triggered by gibberellins during rosette growth in Arabidopsis. Journal of Experimental Botany. 2012;**63**(7): 2769-2786. DOI: 10.1093/jxb/err463

[65] Achard P, Renou JP, Berthome R, Harberd NP, Genschik P. Plant DELLAs restrain growth and promote survival of adversity by reducing the levels of reactive oxygen species. Current Biology. 2008;**18**(9):656-660. DOI: 10.1016/j.cub.2008.04.034

[66] Beneventi MA, da Silva OB, de Sá MEL, et al. Transcription profile of soybean-root-knot nematode interaction reveals a key role of phythormones in the resistance reaction. BMC Genomics. 2013;**14**:322. DOI: 10.1186/1471-2164-14-322

[67] Clark SE. Cell signalling at the shoot meristem. Nature Reviews. Molecular Cell Biology. 2001;**2**:276-284

[68] Siddique S, Radakovic ZS, De La Torre CM, Chronis D, Novák O, Ramireddy E, et al. Proceedings of the National Academy of Sciences. 2015;**112**(41):12669-12674. DOI: 10.1073/pnas.1503657112

[69] Niemann MCE, Bartrina I, Ashikov A, Weber H, Novák O, Spíchal L, et al. Arabidopsis ROCK1 transports UDP-GlcNAc/UDP-GalNAc and regulates ER protein quality control and cytokinin activity. Proceedings of the National Academy of Sciences. 2015;**112**(1):291-296. DOI: 10.1073/pnas.1419050112

[70] Grunewald W, Karimi M, Wieczorek K, Van de Cappelle E, Wischnitzki E, Grundler FM, et al. A role for AtWRKY23in feeding site establishment of plant-parasitic nematodes. Plant Physiology. 2008;**148**(1):358-368

[71] Flores R, Gas M-E, Molina-Serrano D, Nohales M-A, Carbonell A, Gago S, et al. Viroid replication: Rolling-circles, enzymes and ribozymes. Viruses. 2009;**1**:317-334. DOI: 10.3390/v1020317

[72] Hadidi A, Flores R, Randles JW, Semancik JS. Viroids. Collingwood, Australia: CSIRO Publishing; 2003. p. 392

[73] Hammond RW. Columnea latent viroid. In: Hadidi A, Flores R, Randles JW, Semancik JS, editors. Viroids. Collingwood (AU): CSIRO Publishing; 2003. pp. 231-232

[74] Diener TO. Discovering viroids—A person l perspective. Nature Reviews. Microbiology. 2003;**1**:75-80

[75] Clark DP. Viruses, viroids, and prions. In: McGehee MR, editor. Molecular Biology. 3rd ed. London, San Diego, CA: Elsevier/Academic Press; 2019

[76] Mishra MD, Hammond RW, Owens RA, Smith DR, Diener TO. Indian bunchy top disease of tomato plants is caused by a distinct strain of Citrus exocortis viroid (CEVd). Journal of General Virology. 1991;**72**:1-5

[77] Ding B. The biology of viroid-host interactions. Annual Review of Phytopathology. 2009;**47**:105-131

[78] Owens AR, Tech KB, Shao JY, Sano T, Baker J. Global analysis of tomato gene expression during potato spindle tuber viroid infection reveals a complex array of changes affecting hormone signaling. MPMI. 2012;**25**(4):582-598. DOI: 10.1094/MPMI-09-11-0258

[79] Di Serio F, Flores R, ThJ V, Li SF, Pallas V, Randles JW, et al. Current status of viroid taxonomy. Archives of Virology. 2014;**159**:3467-3478

[80] Giguère T, Perreault J-P. Classification of the Pospiviroidae based on their structural hallmarks. PLoS One. 2017;**12**(e0182536)

[81] Kovalskaya N, Hammond RW. Molecular biology of viroid–host interactions and disease control strategies. Plant Science. 2014;**228**:48-60

[82] Gómez G, Martinez G, Pallás V. Interplay between viroid-induced pathogenesis and RNA silencing pathways. Trends in Plant Science. 2009;**14**:264-269

[83] Martin WH. "Spindle tuber", a new potato trouble. Hints to potato growers. 3. New Jersey State Potato Association. 1922;**3**(3):17. II

[84] Singh RP. Experimental host range of the potato spindle tuber 'virus'. American Potato Journal. 1973;**50**: 111-123

[85] Lebas BSM, Elliott DR, Ochoa-Corona FM, Tang J, Alexander BJR. Delimiting survey for potato spindle tuber viroid on tomato and capsicum in New Zealand greenhouses 2003; (abst. 1943)

[86] Daros JA, Flores R. Arabidopsis thaliana has the enzymatic machinery for replicating representative viroid species of the family Pospiviroidae. Proceedings. National Academy of Sciences. United States of America. 2004;**101**:6792-6797

[87] Matousek J, Orctová L, Steger G, Skopek J, Moors M, Dedic P, et al. Analysis of thermal stress-mediated PSTVd variation and biolistic inoculation of progeny of viroid "thermomutants" to tomato and Brassica species. Virology. 2004;**323**:9-23

[88] Itaya A, Zhong X, Bundschuh R, Qi Y, Wang Y, Takeda R, et al. A structured viroid RNA is substrate for Dicer-like cleavage to produce biologically active small RNAs but is resistant to RISC-mediated degradation. Journal of Virology. 2007;**81**:2980-2994

[89] Wang X, Goregaoker SP, Culver JN. Interaction of the tobacco mosaic virus replicase protein with a NAC domain transcription factor is associated with the suppression of systemic host defenses. Journal of Virology. 2009;**83**: 9720-9730

[90] Ascencio-Ibáñez JT, Sozzani R, Lee T-J, Chu T-M, Wolfinger RD, Cella R, et al. The impact of phytohormones on virus infection and disease. Current Opinion in Virology. 2016;**17**:25-31. DOI: 10.1016/j.coviro.2015.11.003

[91] Grant MR, Jones JD. Hormone (dis) harmony moulds plant health and disease. Science. 2009;**324**:750-752

[92] Onaga G, Wydra K. Advances in plant tolerance to biotic stresses. In: Abdurakhmonov IY, editor. Plant Genomics. London: IntechOpen; 2016. pp. 229-272

[93] Zhao S, Li Y. Current understanding of the interplays between host hormones and plant viral infections. PLoS Pathogens. 2021;**17**(2):e1009242. DOI: 10.1371/journal.ppat.1009242

[94] Pieterse CMJ, van der Does D, Zamioudis C, Leon-Reyes A, van Wees SCM. Hormonal modulation of plant immunity. Annual Review of Cell and Developmental Biology. 2012;**28**:489-521

[95] Nakashita H, Yasuda M, Nitta T, Asami T, Fujioka S, Arai Y, et al. Brassinosteroid functions in a broad range of disease resistance in tobacco and rice. The Plant Journal. 2003; **2003**(33):887-898. DOI: 10.1046/j.1365-313x.2003.01675.x

[96] De Bruyne L, Höfte M, De Vleesschauwer D. Connecting growth and defense: The emerging roles of brassinosteroids and gibberellins in plant innate immunity. Molecular Plant. 2014;7:943-959

[97] Robert-Seilaniantz A, Grant M, Jones JDG. Hormone crosstalk in plant disease and defense: More than just jasmonate-salicylate antagonism. Annual Review of Phytopathology. 2011;**49**:317-343

[98] Collum TD, Culver JN. Tobacco mosaic virus infection disproportionatyely impacts phloem associated translatomes in *Arabidopsis thaliana* and *Nicotiana benthamiana*. Virology. 2017;**510**:76-89

[99] Rivas-San Vicente M, Plasencia J. Salicylic acid beyond defence: Its role in plant growth and development. Journal of Experimental Botany. 2011;**62**:3321-3338. DOI: 10.1093/jxb/err031

[100] Vlot AC, Klessig DF, Park S-W. Systemic acquired resistance: The elusive signal(s). Current Opinion in Plant Biology. 2008;**11**:436-442. DOI: 10.1016/j.pbi.2008.05.003

[101] Wani AH, Singh S, Upadhyay N. Salicylic acid to decrease plant stress. Environmental Chemistry Letters. 2017;**15**:101-123

[102] Murphy AM, Carr JP. Salicylic acid has cell-specific effects on tobacco mosaic virus replication and cell-to-cell movement. Plant Physiology. 2002;**128**: 552-563

[103] Durrant WE, Dong X. 2004. Systemic acquired resistance. Annual Review of Phytopathology. 2002;**42**:185-209

[104] Singh A, Lim G-H, Kachroo P. Transport of chemical signals in systemic acquired resistance. Journal of Integrative Plant Biology. 2017;**59**: 336-344. DOI: 10.1111/jipb.12537

[105] Navarre DA, Mayo D. Differential characteristics of salicylic acid-mediated signaling in potato. Physiological and Molecular Plant Pathology. 2004;**64**: 179-188

[106] Ji LH, Ding SW. The suppressor of transgene RNA silencing encoded by cucumber mosaic virus interferes with salicylic asid-mediated virus resistance. Molecular Plant-Microbe Interactions. 2001;**14**(715):724

[107] Baebler S, Stare K, Kovač M, Blejec A, Prezelj N, Stare T, et al. Dynamics of responses in compatible potato-potato virus Y interaction are modulated by salicylic acid. PLoS One. 2011;**6**:e29009

[108] Więsyk A, Iwanicka-Nowicka R, Fogtman A, Zagórski-Ostoja W, Góra-Sochacka A. Time-course microarray analysis reveals differences between transcriptional changes in

tomato leaves triggered by mild and severe variants of potato spindle tuber viroid. Viruses. 2018;**10**:257

[109] Zheng Y, Wang Y, Ding B, Fei Z. Comprehensive transcriptome analyses reveal that potato spindle tuber viroid triggers genome-wide changes in alternative splicing, inducible trans-acting activity of phasiRNAs and immune responses. Journal of Virology. 2017;**91**:e00247-e00217

[110] Wise RP, Moscou MJ, Bogdanove AJ, Whitham SA. Transcript profiling in host-pathogen interactions. Annual Review of Phytopathology. 2007;**45**: 329-369

[111] Belles-Boix E, Hamant O, Witiak SM, Morin H, Traas J, Pautot V. KNAT6: An Arabidopsis homeobox gene involved in meristem activity and organ separation. The Plant Cell. 2006;**2007**:181900-181907

[112] Campos L, Granell P, Tárraga S, López-Gresa P, Conejero V, Bellés JM, et al. Salicylic acid and gen- tisic acid induce RNA silencing-related genes and plant resistance to RNA pathogens. Plant Physiology and Biochemistry. 2014;**77**:35-43

[113] Pokorn T, Radišek S, Javornik B, Štajner N, Jakše J. Development of hop transcriptome to support research into host-Viroid interactions. PLoS One. 2017;**12**:e0184528

[114] Kappagantu M, Bullock JM, Nelson ME, Eastwell KC. Hop stunt viroid: Effect on host (Humulus lupulus) transcriptome and its interactions with hop powdery mildew (*Podspheara macularis*). Molecular Plant-Microbe Interactions. 2017;**30**:842-851

[115] Milanović J, Oklešťková J, Novák O, Mihaljević S. Effects of potato spindle tuber viroid infection on phytohormone and antioxidant responses in symptomless Solanum laxum plants. Journal of Plant Growth Regulation. 2019;**38**:325-332

[116] Milanović J, Oklešťková J, Majdandžić A, Novák O, Mihaljević S. Organ-specific differences in endogenous phyto- hormone and antioxidative responses in potato upon PSTVd infection. Journal of Plant Physiology. 2019;**232**:107-114

[117] López-Gresa MP, Maltese F, Bellés JM, Conejero V, Kim HK, Choi YH, et al. Metabolic response of tomato leaves upon different plant-pathogen interactions. Phytochemical Analysis. 2010;**21**:89-94

[118] Lisón P, Tárraga S, López-Gresa P, Saurí A, Torres C, Campos L, et al. A noncoding plant pathogen provokes both transcriptional and posttranscriptional alterations in tomato. Proteomics. 2013;**13**:833-844

[119] Ahmad P, Rasool S, Gul A, Sheikh SA, Akram NA, Ashraf M, et al. Jasmonates: Multifunctional roles in stress tolerance. Frontiers in Plant Science. 2016;**7**:813

[120] Leon J, Rojo E, Sanchez-Serrano JJ. Wound signalling in plants. Journal of Experimental Botany. 2001;**52**:1-9

[121] Okada K, Abe H, Arimura G. Jasmonates induce both defense responses and communication in monocotyledonous and dicotyledonous plants. Plant & Cell Physiology. 2015;**56**:16-27

[122] Clarke SF, Guy PL, Jameson PE, Schmiere D, Burritt DJ. Influence of white clover mosaic potexvirus infection on the endogenous levels of jasmonic acid and related compound in *Phaseolus vulgaris* L. seedlings. Journal of Plant Physiology. 2000;**156**:433-437

[123] Halim V, Altmann S, Ellinger D, Eschen-Lippold L, Miersch O, Scheel D, et al. PAMP-induced defense responses

in potato required both salicylic acid and jasmonic acid. The Plant Journal. 2009;**57**:230242

[124] Garcia-Marcos A, Pacheco R, Manzano A, Aguilar E, Tenllado F. Oxylipin biosynthesis genes positively regulate programmed cell death during compatible infections with the synergistic pair potato virus X-potato virusY and tomato spotted wilt virus. Journal of Virology. 2013;**87**:5769-5783

[125] Zhu F, Xi DH, Yuan S, Xu F, Zhang DW, Lin HH. Salicylic acid and jasmonic acid are essential for systemic resistance against tobacco mosaic virus in *Nicotiana benthamiana*. Molecular Plant-Microbe Interactions. 2014;**27**:567-577

[126] Milanović J. The role of brassinosteroids and salicylic acid in plant defense response to potato spindle tuber viroid infection [Ph.D. thesis]. University of Zagreb; 2017. Available from: http://bib.irb.hr/datoteka/917706.Jasna_Milanovic_DISSERTATION.pdf

[127] Browse J. Jasmonate passes muster: A receptor and targets for the defense hormone. Annual Review of Plant Biology. 2009;**60**:183-205

[128] Yang D-L, Yao J, Mei C-S, Tong X-H, Zeng L-J, Li Q, et al. Plant hormone jasmonate prioritizes defense over growth by interfering with gibberellin signaling cascade. Proceedings of the National Academy of Sciences of the United States of America. 2012;**109**:E1192-E1200

[129] Grove MD, Spencer GF, Rohwedder WK, Mandava N, Worley JF, Warthen JD Jr, et al. Brassinolide, a plant growth-promoting steroid isolated from Brassica napus pollen. Nature. 1979;**281**:216-217

[130] Choudhary SP, Yu JQ, Yamaguchi-Shinozaki K, Shinozaki K, Tran LS. Benefits of brassinosteroid crosstalk. Trends in Plant Science. 2012;**17**:594-605. DOI: 10.1016/j.tplants.2012.05.012

[131] Ali SS, Gunupuru LR, Kumar GBS, Khan M, Scofield S, Nicholson P, et al. Plant disease resistance is augmented in uzu barley lines modified in the brassinosteroid receptor BRI1. BMC Plant Biology. 2014;**14**:Article number:227

[132] Sharma I, Kaur N, Pati PK. Brassinosteroids: A promising option in deciphering remedial strategies for abiotic stress tolerance in rice. Frontiers in Plant Science. 2017;**8**:2151

[133] Zhang DW, Deng XG, Fu FQ, Lin HH. Induction of plant virus defense response by brassinosteroids and brassinos- teroid signalling in *Arabidopsis thaliana*. Planta. 2015;**241**:875-885

[134] Divi UK, Rahman T, Krishna P. Brassinosteroid-mediated stress tolerance in arabidopsis shows interactions with abscisic acid, ethylene and salicylic acid pathways. BMC Plant Biology. 2010;**10**:151

[135] Lozano-Durán R, Zipfel C. Trade–off between growth and immunity: Role of brassinosteroids. Trends in Plant Science. 2015;**20**:12-19

[136] Doi Y, Teranaka M, Yora K, Asuyama H. Mycoplasma or PLT grouplike microorganisms found in the phloem elements of plants infected with mulberry dwarf, potato witches' broom, aster yellows or paulownia witches' broom. Annals Phytopathological Society of Japan. 1967;**33**:259-266

[137] IRPCM. 'Candidatus Phytoplasma', a taxon for the wall-less, non-helical prokaryotes that colonise plant phloem and insects. International Journal of Systematic and Evolutionary Microbiology. 2004;**54**:1243-1255

[138] Lee I-M, Gundersen-Rindal DE, Davis RE, Bartoszyk IM. Revised

classification scheme of phytoplasmas based on RFLP analyses of 16S rRNA and ribosomal protein gene sequences. International Journal of Systematic and Evolutionary Microbiology. 1998;**48**: 1153-1169

[139] Wei W, Davis RE, Lee I-M, Zhao Y. Computer-simulated RFLP analysis of 16S rRNA genes: Identification of ten new phytoplasma groups. International Journal of Systematic and Evolutionary Microbiology. 2007;**57**:1855-1867

[140] Zhao Y, Wei W, Lee I-M, Shao J, Suo X, Davis RE. Construction of an interactive online phytoplasma classification tool, iPhyClassifier, and its application in analysis of the peach X-disease phytoplasma group (16SrIII). International Journal of Systematic and Evolutionary Microbiology. 2009;**59**: 2582-2593

[141] Marcone C, Neimark H, Ragozzino A, Lauer U, Seemüller E. Chromosome sizes of phytoplasmas composing major phylogenetic groups and subgroups. Phytopathology. 1999;**89**:805-810

[142] Hogenhout SA, Oshima K, Ammar E-D, Kakizawa S, Kingdom HN, Namba S. Phytoplasmas: Bacteria that manipulate plants and insects. Molecular Plant Pathology. 2008;**9**: 403-423

[143] Contaldo N, Satta E, Zambon Y, Paltrinieri S, Bertaccini A. Development and evaluation of different complex media for phytoplasma isolation and growth. Journal of Microbiological Methods. 2016;**127**:105-110. DOI: 10.1016/j.mimet.2016.05.031

[144] Bertaccini A, Lee I-M. Phytoplasmas: An Update. In: Rao GP, Beratccini A, Fiore N, Liefting LW, editors. Phytoplasmas: Plant Pathogenic Bacteria – I. Singapore: Springer Nature Singapore Pte Ltd.; 2018. pp. 1-29. DOI: 10.1007/978-981-13-0119-3_1

[145] Bertaccini A, Bellardi MG, Vibio M. Virus diseases of ornamental shrubs. X. Euphorbiapulcherrima Willd. Infected by viruses and phytoplasmas. Phytopathologia Mediterranea. 1996;**35**:129-132

[146] Lee I-M, Klopmeyer M, Bartoszyk IM, Gundersen-Rindal DE, Chou T, Thomson KL, et al. Phytoplasma induced free-branching in commercial poinsettia cultivars. Nature Biotechnology. 1997;**15**:178-182

[147] Weintraub PG, Beanland LA. Insect vectors of phytoplasmas. Annual Revue of Entomology. 2006;**51**:91-111

[148] Sugio A, Kingdom HN, MacLean AM, Grieve VM, Hogenhout SA. Phytoplasma protein effector SAP11 enhances insect vector reproduction by manipulating plant development and defense hormone biosynthesis. Proceedings of the National Academy of Sciences USA. 2011;**108**:E1254-E1263

[149] Alma A, Bosco D, Danielli A, Bertaccini A, Vibio M, Arzone A. Identification of phytoplasmas in eggs, nymphs and adults of Scaphoideus titanus Ball reared on healthy plants. Insect Molecular Biology. 1997;**6**:115-121

[150] Kawakita H, Saiki T, Wei W, Mitsuhashi W, Watanabe K, Sato M. Identification of mulberry dwarf phytoplasmas in the genital organs and eggs of leafhopper Hishimonoides sellatiformis. Phytopathology. 2000;**90**:909-914

[151] Hanboonsong Y, Choosai C, Panyim S, Damak S. Transovarial transmission of sugarcane white leaf phytoplasma in the insect vector Matsumuratettix hiroglyphicus (Matsumura). Insect Molecular Biology. 2002;**11**:97-103

[152] Tedeschi R, Ferrato V, Rossi J, Alma A. Possible phytoplasma

transovarial transmission in the psyllids *Cacopsylla melanoneura* and *Cacopsylla pruni*. Plant Pathology. 2006;**55**:18-24

[153] Khan AJ, Botti S, Paltrinieri S, Al-Subhi AM, Bertaccini A. Phytoplasmas in alfalfa seedlings: Infected or contaminated seeds? In: 13th Congress International Organization for Mycoplasmoloy; Vienna, Austria. Singapore: Springer; 2002. p. 6

[154] Botti S, Bertaccini A. Phytoplasma infection trough seed transmission: further observations. In: 16th Congress International Organization for Mycoplasmology, Cambridge, United Kingdom. Vol. 76. 2006. p. 113

[155] Calari A, Paltrinieri S, Contaldo N, Sakalieva D, Mori N, Duduk B, et al. Molecular evidence of phytoplasmas in winter oilseed rape, tomato and corn seedlings. Bulletin of Insectology. 2011;**64**:S157-S158

[156] Chung BN, Jeong MI. Identification of "stolbur" phytoplasmas in Petunia hybdrida seedlings. Phytopathogenic Mollicutes. 2014;**4**:5-8

[157] Satta E, Contaldo N, Paltrinieri S, Bertaccini A. Biological and molecular proof of phytoplasma seed transmission in corn. In: 21th Congress of the International Organization for Mycoplasmology; Brisbane, Australia. Vol. 61. 2016. pp. 65-66

[158] Bertaccini A, Davis RE, Lee I-M. In vitro micropropagation for maintenance of mycoplasmalike organisms in infected plant tissues. Horticultural Science. 1992;**27**:1041-1043

[159] Jarausch W, Lansac M, Dosba F. Long-term maintenance of nonculturable apple-proliferation phytoplasmas in their micropropagated natural host plant. Plant Pathology. 1996;**45**:778-786

[160] Bertaccini A. Phytoplasmas: Diversity, taxonomy, and epidemiology. Frontiers in Bioscience. 2007;**12**:673-689

[161] Dermastia M. Plant hormones in phytoplasma infected plants. Frontiers in Plant Science. 2019;**10**(477):1-15. DOI: 10.3389/fpls.2019.00477

[162] Rijavec T, Dermastia M. Cytokinins and their function in developing seeds. Acta Chimica Slovenica. 2010;**57**:617-629

[163] Kohli A, Sreenivasulu N, Lakshmanan P, Kumar PP. The phytohormone crosstalk paradigm takes center stage in understanding how plants respond to abiotic stresses. Plant Cell Reports. 2013;**32**:945-957. DOI: 10.1007/s00299-013-1461-y

[164] Wasternack C, Song SJ. Biosynthesis, metabolism, and signaling by proteins activating and repressing transciption. Journal of Experimental Botany. 2016;**68**(erw443). DOI: 10.1093/jxb/erw443

[165] Liu J, Moore S, Chen C, Lindsey K. Crosstalk complexities between auxin, cytokinin, and ethylene in Arabidopsis root development: From experiments to systems modeling, and back again. Molecular Plant. 2017;**10**:1480-1496. DOI: 10.1016/J.MOLP.2017.11.002

[166] Ahanger MA, Ashraf M, Bajguz A, Ahmad P. Brassinosteroids regulate growth in plants under stressful environments and crosstalk with other potential phytohormones. Journal of Plant Growth Regulation. 2018;**37**:1007-1024. DOI: 10.1007/s00344-018-9855-2

[167] Binenbaum J, Weinstain R, Shani E. Gibberellin localization and transport in plants. Trends in Plant Science. 2018;**23**:410-421. DOI: 10.1016/J.TPLANTS.2018.02.005

[168] Camara MC, Vandenberghe LPS, Rodrigues C, de Oliveira J, Faulds C, Bertrand E. Current advances in gibberellic acid (GA3[1]). Production, patented technologies and potential applications. Planta. 2018;**248**:1049-1062. DOI: 10.1007/s00425-018-2959-x

[169] Ma Y, Cao J, He J, Chen Q, Li X, Yang Y. Molecular mechanism for the regulation of ABA homeostasis during plant development and stress responses. International Journal of Molecular Sciences. 2018;**19**:3643. DOI: 10.3390/ijms19113643

[170] Skalický V, Kubeš M, Napier R, Novák O. Auxins and cytokinins—The role of subcellular organization on homeostasis. International Journal of Molecular Sciences. 2018;**19**:3115. DOI: 10.3390/ijms19103115

[171] Weijers D, Nemhauser J, Yang Z. Auxin: Small molecule, big impact. Journal of Experimental Botany. 2018;**69**:133-136. DOI: 10.1093/jxb/erx463

[172] Wasternack C, Strnad M. Jasmonates are signals in the biosynthesis of secondary metabolites—Pathways, transcription factors and applied aspects—A brief review. New Biotechnology. 2019;**48**:1-11. DOI: 10.1016/j.nbt.2017.09.007

[173] Shigenaga AM, Argueso CT. No hormone to rule them all: Interactions of plant hormones during the responses of plants to pathogens. Seminars in Cell & Developmental Biology. 2016;**56**:174-189. DOI: 10.1016/J.SEMCDB.2016.06.005

[174] Shigenaga AM, Berens ML, Tsuda K, Argueso CT. Towards engineering of hormonal crosstalk in plant immunity. Current Opinion in Plant Biology. 2017;**38**:164-172. DOI: 10.1016/j.pbi.2017.04.021

[175] Karasov TL, Chae E, Herman JJ, Bergelson J. Mechanisms to mitigate the trade-off between growth and defense. Plant Cell. 2017;**29**:666-680. DOI: 10.1105/tpc.16.00931

[176] van Loon LC, Rep M, Pieterse CMJ. Significance of inducible defense-related proteins in infected plants. Annual Review of Phytopathology. 2006;**44**:135-162. DOI: 10.1146/annurev.phyto.44.070505.143425

[177] F. Acetylsalicylic acid (aspirin) induces resistance to tobacco mosaic virus in tobacco. Virology. 1979;**99**:410-412

[178] Fu ZQ, Dong X. Systemic acquired resistance: Turning local infection into global defense. Annual Review of Plant Biology. 2013;**64**:839-863. DOI: 10.1146/annurev-arplant-042811-105606

[179] Frías M, Brito N, González C. The Botrytis cinerea cerato-platanin BcSpl1 is a potent inducer of systemic acquired resistance (SAR) in tobacco and generates a wave of salicylic acid expanding from the site of application. Molecular Plant Pathology. 2013;**14**:191-196. DOI: 10.1111/j.1364-3703.2012.00842.x

[180] Zhong B-X, Shen Y-W. Accumulation of pathogenesis-related type-5 like proteins in phytoplasma-infected garland chrysanthemum chrysanthemum coronarium. Acta Biochimica et Biophysica Sinica. 2004;**36**:773-779. DOI: 10.1093/abbs/36.11.773

[181] Monavarfeshani A, Mirzaei M, Sarhadi E, Amirkhani A, Khayam Nekouei M, Haynes PA, et al. Shotgun proteomic analysis of the Mexican lime tree infected with 'Candidatus Phytoplasma aurantifolia'. Journal of Proteome Research. 2013;**12**:785-795. DOI: 10.1021/pr300865t

[182] Gai Y-P, Yuan S-S, Liu Z-Y, Zhao H-N, Liu Q, Qin R-L, et al. Integrated phloem sap mRNA and protein expression analysis reveals phytoplasma infection responses in mulberry. Molecular & Cellular Proteomics. 2018;**17**:1702-1719. DOI: 10.1074/mcp.RA118.000670

[183] Hren M, Nikoli'c P, Rotter A, Blejec A, Terrier N, Ravnikar M, et al. "Bois noir" phytoplasma induces

significant reprogramming of the leaf transcriptome in the field grown grapevine. BMC Genomics. 2009;**10**:460. DOI: 10.1186/1471-2164-10-460

[184] Dermastia M, Nikolic P, Chersicola M, Gruden K. Transcriptional profiling in infected and recovered grapevine plant responses to 'Candidatus Phytoplasma solani'. Phytopathogenic Mollicutes. 2015;**5**:S123. DOI: 10.5958/2249-4677.2015.00053.5

[185] Prezelj N, Fragener L, Weckwerth W, Dermastia M. Metabolome of grapevine leaf vein-enriched tissue infected with 'Candidatus Phytoplasma solani'. Mitteilungen Klosterneubg Rebe und Wein Obs und Früchteverwertung. 2016;**66**:74-78

[186] Paolacci AR, Catarcione G, Ederli L, Zadra C, Pasqualini S, Badiani M, et al. Jasmonate-mediated defence responses, unlike salicylate-mediated responses, are involved in the recovery of grapevine from bois noir disease. BMC Plant Biology. 2017;**17**:118. DOI: 10.1186/s12870-017-1069-4

[187] Rotter A, Nikoli'c P, Turnšek N, Kogovšek P, Blejec A, Gruden K, et al. Statistical modeling of long-term grapevine response to 'Candidatus Phytoplasma solani' infection in the field. European Journal of Plant Pathology. 2018;**150**:653-668. DOI: 10.1007/s10658-017-1310-x

[188] Yan C, Xie D. Jasmonate in plant defence: sentinel or double agent? Plant Biotechnology Journal. 2015;**13**:1233-1240. DOI: 10.1111/pbi.12417

[189] Wasternack C. How jasmonates earned their laurels: Past and present. Journal of Plant Growth Regulation. 2015;**34**:761-794. DOI: 10.1007/s00344-015-9526-5

[190] Nejat N, Cahill DM, Vadamalai G, Ziemann M, Rookes J, Naderali N. Transcriptomics-based analysis using RNA-Seq of the coconut (Cocos nucifera) leaf in response to yellow decline phytoplasma infection. Molecular Genetics and Genomics. 2015;**290**:1899-1910. DOI: 10.1007/s00438-015-1046-2

[191] Iqbal N, Khan NA, Ferrante A, Trivellini A, Francini A, Khan MIR. Ethylene role in plant growth, development and senescence: Interaction with other phytohormones. Frontiers in Plant Science. 2017;**8**:475. DOI: 10.3389/fpls.2017.00475

[192] Fan G, Xu E, Deng M, Zhao Z, Niu S. Phenylpropanoid metabolism, hormone biosynthesis and signal transduction-related genes play crucial roles in the resistance of Paulownia fortunei to paulownia witches' broom phytoplasma infection. Genes Genomics. 2015;**37**:913-929. DOI: 10.1007/s13258-015-0321-2

[193] Mardi M, Karimi Farsad L, Gharechahi J, Salekdeh GH. In-depth transcriptome sequencing of Mexican lime trees infected with 'Candidatus Phytoplasma aurantifolia'. PLoS One. 2015;**10**:e0130425. DOI: 10.1371/journal.pone.0130425

[194] Shao F, Zhang Q, Liu H, Lu S, Qiu D. Genome-wide identification and analysis of microRNAs involved in witches'-broom phytoplasma response in Ziziphus jujuba. PLoS One. 2016;**11**:e0166099. DOI: 10.1371/journal.pone.0166099

[195] Snyman MC, Solofoharivelo M-C, Souza-Richards R, Stephan D, Murray S, Burger JT. The use of high-throughput small RNA sequencing reveals differentially expressed microRNAs in response to aster yellows phytoplasma-infection in Vitis vinifera cv. "Chardonnay". PLoS One. 2017;**12**:e0182629. DOI: 10.1371/journal.pone.0182629

[196] Kunkel BN, Harper CP. The roles of auxin during interactions between

bacterial plant pathogens and their hosts. Journal of Experimental Botany. 2018;**69**:245-254. DOI: 10.1093/jxb/erx447

[197] Ehya F, Monavarfeshani A, Mohseni Fard E, Karimi Farsad L, Khayam Nekouei M, Mardi M. Phytoplasma-responsive microRNAs modulate hormonal, nutritional, and stress signalling pathways in Mexican lime trees. PLoS One. 2013;**8**:66372. DOI: 10.1371/journal.pone.0066372

[198] Davey JE, Van Staden J, De Leeuw GTN. Endogenous cytokinin levels and development of flower virescence in Catharanthus roseus infected with mycoplasmas. Physiological Plant Pathology. 1981;**19**: 193-200. DOI: 10.1016/S0048-4059(81) 80021-5

[199] Jones JD, Dangl JL. The plant immune system. Nature. 2006;**444**: 323-329

[200] Chisholm ST, Coaker G, Day B, Staskawicz BJ. Host–microbe interactions: Shaping the evolution of the plant immune response. Cell. 2006;**124**:803-814

[201] Spaink HP. Root nodulation and infection factors produced by Rhizobial. Bacteria. 2000;**54**:257-288. DOI: 10.1146/annurev.micro.54.1.257

[202] Zamioudis C, Pieterse CMJ. Modulation of host immunity by. Beneficial Microbes. Mol Plant Microbe Interact. 2012;**25**(2):139-150. DOI: 10.1094/MPMI-06-11-0179

[203] Harrison MJ. Signaling in the Arbuscular Mycorrhizal Symbiosis. Annual Review of Microbiology. 2005;**59**:19-42. DOI: 10.1146/annurev.micro.58.030603.123749

[204] Kloepper JW, Ryu C-M, Zhang SA. Induced systemic resistance and promotion of plant growth by Bacillus spp. Phytopathology. 2004;**94**: 1259-1266

[205] Lugtenberg B, Kamilova F. Plant-growth-promoting rhizobacteria. Annual Review of Microbiology. 2009;**63**:541-556

[206] Khalimi K, Wirya G. Pemanfaatan Plant Growth Promoting Rhizobacteria Untuk Biostimulants Dan Bioprotectants. Journal of Environmental Science. 2009;**4**(2):131-135

[207] Yadav J, Verma JP, Tiwari KN. Effect of plant growth promoting Rhizobacteria on seed germination and plant growth chickpea (*Cicer arietinum* L.) under in vitro conditions. Biological Forum—An International Journal. 2010;**2**(2):15-18

[208] Kloepper JW, Lifshitz R, Zablotowicz RM. Freeliving bacteria inocula for enhancing crop productivity. Trends Biotechnolology. 1989;**7**:39-44

[209] Chanway CP, Holl FB. First year yield performance of spruce seedlings inoculated with plant growth promoting rhizobacteria. Canadian Journal of Microbiology. 1993;**39**:1084-1088

[210] Wang W, Vinocur B, Altman A. Plant responses to drought, salinity and extreme temperatures: Towards genetic engineering for stress tolerance. Planta. 2003;**218**(1):1-14

[211] Karadeniz A, Topeuoglu SF, Inan S. Auxin, gibberellin, cytokinin and abscisic acid production in some bacteria. World Journal of Microbiology and Biotechnology. 2006;**22**:1061-1064

[212] Vlot CA, Dempsey DA, Klessig DF. Salicylic acid, a multifaceted hormone to Combat Disease. Annual Review of Phytopathology. 2009;**47**:177-206. DOI: 10.1146/annurev.phyto.050908.135202

[213] Glazebrook J. Contrasting mechanisms of defense against

biotrophic and necrotrophic pathogens. Annual Review of Phytopathology. 2005;**43**:205-227

[214] Klessig DF, Malamy J. The salicylic acid signal in plants. Plant Molecular Biology. 1994;**26**:1439-1458

[215] Enyedi AJ, Yalpani N, Silverman P, Raskin I. Signal molecules in systemic plant resistance to pathogens and pests. Cell. 1992;**70**:879-886

[216] Clarke JD, Volko SM, Ledford H, Ausubel FM, Dong X. Roles of salicylic acid, jasmonic acid, and ethylene in cpr-induced resistance in Arabidopsis. Plant Cell. 2000;**12**:2175-2190

[217] Wang C, Avdiushko S, Hildebrand D. Overexpression of a cytoplasm-localized allene oxide synthase promotes the wound-induced accumulation of jasmonic acid in transgenic tobacco. Plant Molecular Biology. 1999;**40**:783-793

[218] Pozo MJ, Van Loon LC, Pieterse CMJ. Jasmonates—Signals in plant microbe interactions. Journal of Plant Growth Regulation. 2004;**23**:211-222

[219] Bell E, Mullet JE. Lipoxygenase gene expression is modulated in plants by water deficit, wounding, and methyl jasmonate. Molecular Genetic and Genomics. 1991;**230**:456-462

[220] Melan MA, Dong X, Endara ME, Davis KR, Ausubel FM, Peterman TK. An Arubidopsis thulium lipoxygenase gene is induced by pathogens, abscisic acid, and methyl jasmonate. Plant Physiology. 1993;**101**:441-450

[221] Laudert D, Weiler EW. Allene oxide synthase: A major control point in Arabidopsis thaliana octadecanoid signalling. The Plant Journal. 1998;**15**(5):675-684. DOI: 10.1046/j.1365-313x.1998.00245.x

[222] Heitz T, Bergey DR, Ryan CA. A gene encoding achloroplast-targeted lipoxygenase in tomato leaves istransiently induced by wounding, systemin, and methyljasmonate. Plant Physiology. 1997;**114**:1085-1093

[223] Ueda J, Kato J. Identification of jasmonic acid and abscisic acid as senescence-promoting substances from Cleyera ochnacea DC. Agricultural and Biological Chemistry. 1982;**46**:1975-1976

[224] Pauw B, Memelink J. Jasmonate-responsive gene expression. Journal of Plant Growth Regulation. 2004;**23**: 200-210

[225] van Loon LC, Rep M, Pieterse CM. Significance of inducible defence-related proteins in infected plants. Annual Review of Phytopathology. 2006;**44**:135-162

[226] Broekaert WF, Delaure SL, De Bolle MF, Cammue BP. The role of ethylene in host-pathogen interactions. Annual Review of Phytopathology. 2006;**44**:393-416

[227] Brodersen P, Petersen M. Arabidopsis MAP kinase 4 regulates salicylic acid- and jasmonic acid/ethylene-dependent responses via EDS1 and PAD4. The Plant Journal. 2006;**47**:532-546

[228] Von Dahl CC, Baldwin IT. Deciphering the role of ethylene in plant-herbivore interactions. Journal of Plant Growth Regulation. 2007;**26**: 201-209

[229] Sharp RE. Interaction with ethylene: Changing views on the role of abscisic acid in root and shoot growth responses to water stress. Plant Cell Environment. 2002;**25**:211-222. DOI: 10.1046/j.1365-3040.2002.00798.x

[230] Cutler SR, Rodriguez PL, Finkelstein RR, Abrams SR. Abscisic acid: Emergence of a core signaling network. Annual Review of Plant Biology. 2010;**61**:651-679

[231] Cao F, Yoshioka K, Desveaux D. The roles of ABA in plantpathogen interactions. Journal of Plant Research. 2011;**124**:489-499

[232] Ton J, Flors V, Mauch-Mani B. The multifaceted role of ABA in disease resistance. Trends in Plant Science. 2009;**14**:310-317

[233] de Torres-Zabala M, Truman W, Bennett MH, Lafforgue G, Mansfield JW, Rodriguez EP, et al. Pseudomonas syringae pv. Tomato hijacks the Arabidopsis abscisic acid signalling pathway to cause disease. The EMBO Journal. 2007;**26**:1434-1443

[234] Jiang K, Sorefan K, Deeks MJ, Bevan MW, Hussey PJ, Hetherington AM. The ARP2/3 complex mediates guard cell actin reorganization and stomatal movement in Arabidopsis. The Plant Cell. 2012;**24**:2031-2040

[235] Yasuda M, Ishikawa A, Jikumaru Y, Seki M, Umezawa T, Asami T, et al. Antagonistic interaction between systemic acquired resistance and the abscisic acid–mediated abiotic stress response in Arabidopsis. The Plant Cell. 2008;**20**(6):1678-1692

[236] Traens L. The role of the PPR protein PGN in the resistance against biotic and abiotic stress [Master thesis, Ghent University], Ghent. 2009. p. 1-69

[237] Shi F, Dong Y, Zhang Y, Yang X, Qiu D. Overexpression of the PeaT1 elicitor gene from Alternaria tenuissima improves drought tolerance in rice plants via interaction with a myo-inositol oxygenase. Frontiers in Plant Science. 2017;**8**:970

[238] Hasanuzzaman M, Nahar K, Alam MM, Ahmad S, Fujita M. Exogenous application of phytoprotectants in legumes against environmental stress. In: Azooz MM, Ahmed P, editors. Legumes under Environmental Stress. John Wiley & Sons, Ltd.; 2015. pp. 161-197

[239] Chen Z, Pottosin II, Cuin TA, Fuglsang AT, Tester M, Jha D, et al. Root plasma membrane transporters controlling K+/Na+ homeostasis in salt-stressed barley. Plant Physiology. 2007;**145**(4):1714-1725

[240] Navarro L, Bari R, Seilaniantz A, Nemri A, Jones JD. Roles of plant hormones in plant resistance and susceptibility to pathogens. In: Genomics of Disease. New York, NY: Springer; 2008. pp. 1-10

[241] Jacobs GSS, Du G. Efficacy, rate and timing of applications of abscisic acid to enhance fruit anthocyanin contents in 'cabernet Sauvignon'grapes. The Journal of Horticultural Science and Biotechnology. 2011;**86**(5):505-510

[242] Millet YA, Danna CH, Clay NK, Songnuan W, Simon MD, Werck-Reichhart D, et al. Innate immune responses activated in Arabidopsis roots by microbe-associated molecular patterns. The Plant Cell. 2010;**22**(3):973-990

[243] Zamioudis C, Pieterse CM. Modulation of host immunity by beneficial microbes. Molecular Plant-Microbe Interactions. 2012;**25**(2): 139-150

[244] Doornbos RF, Geraats BP, Kuramae EE, Van Loon LC, Bakker PA. Effects of jasmonic acid, ethylene, and salicylic acid signaling on the rhizosphere bacterial community of Arabidopsis thaliana. Molecular Plant-Microbe Interactions. 2011;**24**(4):395-407

[245] Berendsen RL, Pieterse CM, Bakker PA. The rhizosphere microbiome and plant health. Trends in Plant Science. 2012;**17**(8):478-486

[246] Chandler JW, Cole M, Jacobs B, Comelli P, Werr W. Genetic integration of DORNRÖSCHEN and DORNRÖSCHEN-LIKE reveals hierarchical interactions in auxin

signalling and patterning of the Arabidopsis apical embryo. Plant Molecular Biology. 2011;**75**(3):223-236

[247] Achard P, Genschik P. Releasing the brakes of plant growth: How GAs shutdown DELLA proteins. Journal of Experimental Botany. 2009;**60**(4): 1085-1092

[248] De Vleesschauwer D, Gheysen G, Höfte M. Hormone defense networking in rice: Tales from a different world. Trends in Plant Science. 2013;**18**(10): 555-565

[249] Bakker PA, Berendsen RL, Van Pelt JA, Vismans G, Yu K, Li E, et al. The soil-borne identity and microbiome-assisted agriculture: Looking back to the future. Molecular Plant. 2020;**13**(10): 1394-1401

[250] Van der Ent S, Van Wees SC, Pieterse CM. Jasmonate signaling in plant interactions with resistance-inducing beneficial microbes. Phytochemistry. 2009;**70**(13-14):1581-1588

[251] Alazem M, Lin NS. Roles of plant hormones in the regulation of host–virus interactions. Molecular Plant Pathology. 2015;**16**(5):529-540

[252] Bulgarelli D, Schlaeppi K, Van Spaepen S, Themaat EVL, Schulze-Lefert P. Structure and functions of the bacterial microbiota of plants. Annual Review of Plant Biology. 2013;**64**: 807-838

[253] Reinhold-Hurek B, Bünger W, Burbano CS, Sabale M, Hurek T. Roots shaping their microbiome: Global hotspots for microbial activity. Annual Review of Phytopathology. 2015;**53**:403-424

[254] Hussain A, Ullah I, Hasnain S. Microbial manipulation of auxins and cytokinins in plants. In: Auxins and Cytokinins in Plant Biology. New York, NY: Humana Press; 2017. pp. 61-72

[255] Wang SD, Zhu F, Yuan S, Yang H, Xu F, Shang J, et al. The roles of ascorbic acid and glutathione in symptom alleviation to SA-deficient plants infected with RNA viruses. Planta. 2011;**234**(1):171-181

[256] Takahashi H, Miller J, Nozaki Y, Sukamto, Takeda M, Shah J, et al. RCY1, an Arabidopsis thaliana RPP8/HRT family resistance gene, conferring resistance to cucumber mosaic virus requires salicylic acid, ethylene and a novel signal transduction mechanism. The Plant Journal. 2002;**32**(5):655-667

[257] Zhu T, Zhou X, Zhang JL, Zhang WH, Zhang LP, You CX, et al. Ethylene-induced NbMYB4L is involved in resistance against tobacco mosaic virus in *Nicotiana benthamiana*. Molecular Plant Pathology. 2022;**23**(1):16-31. DOI: 10.1111/ mpp.13139

[258] Yasuda M, Ishikawa A, Jikumaru Y, Seki M, Umezawa T, Asami T, et al. Antagonistic interaction between systemic acquired resistance and the abscisic acid-mediated abiotic stress response in Arabidopsis. Plant Cell. 2008;**20**(6):1678-1692. DOI: 10.1105/ tpc.107.054296. Epub 2008 Jun 27

[259] Sawinski K, Mersmann S, Robatzek S, Böhmer M. Guarding the green: Pathways to stomatal immunity. Molecular Plant-Microbe Interactions. 2013;**26**:626-632

[260] Chivasa S, Murphy AM, Naylor CJP. Salicylic acid interferes with tobacco mosaic virus replication via a novel salicylhydroxamic acid-sensitive mechanism. The Plant Cell. 1997;**9**(4):547-557

[261] Ji LH, Ding SW. The suppressor of transgene RNA silencing encoded by cucumber mosaic virus interferes with salicylic acid-mediated virus resistance. Molecular Plant-Microbe Interactions. 2001;**14**(6):715-724

[262] Gilliland A, Singh DP, Hayward JM, Moore CA, Murphy AM, York CJ, et al. Genetic modification of alternative respiration has differential effects on antimycin A-induced versus salicylic acid-induced resistance to tobacco mosaic virus. Plant Physiology. 2003;**132**:1518-1152

[263] Liao YWK, Liu YR, Liang JY, Wang WP, Zhou J, Xia XJ, et al. The relationship between the plant-encoded RNA-dependent RNA polymerase 1 and alternative oxidase in tomato basal defense against Tobacco mosaic virus. Planta. 2015;**241**(3):641-650

[264] Greve K, La Cour T, Jensen MK, Poulsen FM, Skriver K. Interactions between plant RING-H2 and plant-specific NAC (NAM/ATAF1/1/CUC2) proteins: RING-H2 molecular specificity and cellular localization. The Biochemical Journal. 2003;**371**:97-108

[265] Huh SU, Lee GJ, Kim YJ, Paek KH. Capsicum annuum WRKY transcription factor d (CaWRKYd) regulates hypersensitive response and defense response upon Tobacco mosaic virus infection. Plant Science. 2012;**197**:50-58

[266] Whenham RJ, Fraser RSS, Brown LP, Payne JA. Tobacco mosaic virus-induced increase in abscisic acid concentration in tobacco leaves: Intracellular location in light and dark green areas, and relationship to symptomdevelopment. Planta. 1986;**168**(4):592-598

[267] Chen L, Zhang L, Li D, Wang F, Yu D. WRKY8 transcription factor functions in the TMV-cg defense response by mediating both abscisic acid and ethylene signaling in Arabidopsis. Proceedings of the National Academy of Sciences. 2013;**110**(21):E1963-E1971

[268] Iriti M, Faoro F. Abscisic acid is involved in chitosan-induced resistance to tobacco necrosis virus (TNV). Plant Physiology and Biochemistry. 2008;**46**(12):1106-1111

[269] Mauch-Mani B, Mauch F. The role of abscisic acid in plant-pathogen interactions. Current Opinion in Plant Biology. 2005;**8**(4):409-414

[270] Ambawat S, Sharma P, Yadav NR, Yadav RC. MYB transcription factor genes as regulators for plant responses: An overview. Physiology and Molecular Biology of Plants. 2013;**19**(3):307-321

[271] Seo PJ, Park CM. MYB96-mediated abscisic acid signals induce pathogen resistance response by promoting salicylic acid biosynthesis in Arabidopsis. New Phytologist. 2010;**186**(2):471-483

[272] Zhu L, Li Y, Li L, Yang J, Zhang M. Ethylene is involved in leafy mustard systemic resistance to Turnip mosaic virus infection through the mitochondrial alternative oxidase pathway. Physiological and Molecular Plant Pathology. 2011;**76**(3-4):166-172

[273] Knoester M, Linthorst HJM, Bol JF, Van Loon LC. Involvement of ethylene in lesion development and systemic acquired resistance in tobacco during the hypersensitive reaction to tobacco mosaic virus. Physiological and Molecular Plant Pathology. 2001;**59**:45-57

[274] Fischer U, Dröge-Laser W. Overexpression of NtERF5, a new member of the tobacco ethylene response transcription factor family enhances resistance to tobacco mosaic virus. Molecular Plant-Microbe Interactions. 2004;**17**(10):1162-1171

[275] Pumplin N, Voinnet O. RNA silencing suppression by plant pathogens: Defence, counter-defence and counter-counter-defence. Nature Reviews. Microbiology. 2013;**11**:745-760

[276] Lewsey MG, Murphy AM, Maclean D, Dalchau N, Westwood JH, Macaulay K, et al. Disruption of two defensive signaling pathways by a viral RNA silencing suppressor. Molecular

Plant-Microbe Interactions. 2010;**23**: 835-845

[277] Diaz-Pendon JA, Li F, Li WX, Ding SW. Suppression of antiviral silencing by cucumber mosaic virus 2b protein in Arabidopsis is associated with drastically reduced accumulation of three classes of viral small interfering RNAs. Plant Cell. 2007;**19**:2053-2063

[278] Vidhyasekaran P. Salicylic acid signaling in plant innate immunity. In: Plant Hormone Signaling Systems in Plant Innate Immunity. Dordrecht: Springer; 2015. pp. 27-122

[279] Jovel J, Walker M, H. Sanfacon salicylic acid-dependent restriction of tomato ringspot virus spread in tobacco is accompanied by a hypersensitive response, local RNA silencing, and moderate systemic resistance. Molecular Plant-Microbe Interactions. 2011;**24**: 706-718

[280] Alamillo JM, Saenz P, Garcıa JA. Salicylic acidmediated and RNA-silencing defence mechanisms cooperate in the restriction of systemic spread of plum pox virus in tobacco. The Plant Journal. 2006;**48**:217-227

[281] Yang L, Meng D, Wang Y, Wu Y, Lang C, Jin T, et al. The viral suppressor HCPro decreases DNA methylation and activates auxin biosynthesis genes. Virology. 2020;**546**:133-140

[282] Nie X. Salicylic acid suppresses potato virus Y isolate N:O-induced symptoms in tobacco plants. Phytopathology. 2006;**96**(3):255-263. DOI: 10.1094/PHYTO-96-0255

[283] Leibman D, Wolf D, Saharan V, Zelcer A, Arazi T, Yoel S, et al. A high level of transgenic viral small RNA is associated with broad potyvirus resistance in cucurbits. Molecular Plant-Microbe Interactions. 2011;**24**: 1220-1238

[284] Yang SJ, Carter SA, Cole AB, Cheng NH, Nelson RS. A natural variant of a host RNA-dependent RNA polymerase is associated with increased susceptibility to viruses by Nicotiana benthamiana. Proceedings of the National Academy of Sciences of the United States of America. 2004;**101**:6297-6302

[285] Kazan K, Manners JM. Linking development to defense: auxin in plant-pathogen interactions. Trends in Plant Science. 2009;**14**(7):373-382. DOI: 10.1016/j.tplants.2009.04.005

[286] Padmanabhan MS, Kramer SR, Wang X, Culver JN. Tobacco mosaic virus replicase-auxin/indole acetic acid protein interactions: Reprogramming the auxin response pathway to enhance virus infection. Journal of Virology. 2008;**82**(5):2477-2485

[287] Zhang H, Tan X, Li L, He Y, Hong G, Li J, et al. Suppression of auxin signalling promotes rice susceptibility to rice black streaked dwarf virus infection. Molecular Plant Pathology. 2019;**20**(8):1093-1104

[288] Zhang H, Li L, He Y, Qin Q, Chen C, Wei Z, et al. Distinct modes of manipulation of rice auxin response factor OsARF17 by different plant RNA viruses for infection. Proceedings of the National Academy of Sciences. 2020;**117**(16):9112-9121

[289] Collum TD, Padmanabhan MS, Hsieh YC, Culver JN. Tobacco mosaic virus-directed reprogramming of auxin/indole acetic acid protein transcriptional responses enhances virus phloem loading. Proceedings of the National Academy of Sciences. 2016;**113**(19): E2740-E2749

[290] Zhai Y, Peng H, Neff MM, Pappu HR. Putative auxin and light responsive promoter elements from the Tomato spotted wilt tospovirus genome, when expressed as cDNA, are functional in Arabidopsis. Frontiers in Plant Science. 2019;**10**:804

[291] Zhu S, Gao F, Cao X, Chen M, Ye G, Wei C, et al. The rice dwarf virus P2 protein interacts with ent-kaurene oxidases in vivo, leading to reduced biosynthesis of gibberellins and rice dwarf symptoms. Plant Physiology. 2005;**139**:1935-1194

[292] Kuriger WE, Agrios GN. Cytokinin levels and kinetin-virus interactions in tobacco ringspot virus infected cowpea plants. Phytopathology. 1977;**67**:604-609

[293] Pogány M, Koehl J, Heiser I, Elstner EF, Barna B. Juvenility of tobacco induced by cytokinin gene introduction decreases susceptibility to Tobacco necrosis virus and confers tolerance to oxidative stress. Physiological and Molecular Plant Pathology. 2004;**65**(1):39-47

[294] Spoustová P, Hýsková V, Müller K, Schnablová R, Ryšlavá H, Čeřovská N, et al. Tobacco susceptibility to Potato virus YNTN infection is affected by grafting and endogenous cytokinin content. Plant Science. 2015;**235**:25-36

[295] Selman IW, Arulpragasam PV. Response of tomato plants infected with cucumber mosaic virus to foliar sprays of gibberellic acid. Annals of Botany. 1970;**34**(5):1107-1114

[296] Miozzi L, Napoli C, Sardo L, Accotto GP. Transcriptomics of the interaction between the monopartite phloem-limited geminivirus tomato yellow leaf curl Sardinia virus and Solanum lycopersicum highlights a role for plant hormones, autophagy and plant immune system fine tuning during infection. PLoS One. 2014;**9**:e89951. DOI: 10.1371/journal.pone.0089951

[297] Nakashita H, Yasuda M, Nitta T, Asami T, Fujioka S, Arai Y, et al. Brassinosteroid functions in a broad range of disease resistance in tobacco and rice. The Plant Journal. 2003;**33**(5): 887-898. DOI: 10.1046/j.1365-313x. 2003.01675.x

[298] Sofy AR, Dawoud RA, Sofy MR, Mohamed HI, Hmed AA, El-Dougdoug NK. Improving regulation of enzymatic and non-enzymatic antioxidants and stress-related gene stimulation in Cucumber mosaic cucumovirus-infected cucumber plants treated with glycine betaine, chitosan and combination. Molecules. 2020;**25**(10):2341

[299] Hu J, Huang J, Xu H, Wang Y, Li C, Wen P, et al. Rice stripe virus suppresses jasmonic acid-mediated resistance by hijacking brassinosteroid signaling pathway in rice. PLoS Pathogens. 2020;**16**(8):e1008801. DOI: 10.1371/journal.ppat.1008801